Humber College Library
3199 Lakeshore Blvd. West
Toronto, ON   M8V 1K8

Water in Textiles and Fashion

Water in Textiles and Fashion

Consumption, Footprint, and Life Cycle Assessment

Edited by

Subramanian Senthilkannan Muthu

Woodhead Publishing is an imprint of Elsevier
The Officers' Mess Business Centre, Royston Road, Duxford, CB22 4QH, United Kingdom
50 Hampshire Street, 5th Floor, Cambridge, MA 02139, United States
The Boulevard, Langford Lane, Kidlington, OX5 1GB, United Kingdom

Copyright © 2019 Elsevier Ltd. All rights reserved.

No part of this publication may be reproduced or transmitted in any form or by any means, electronic or mechanical, including photocopying, recording, or any information storage and retrieval system, without permission in writing from the publisher. Details on how to seek permission, further information about the Publisher's permissions policies and our arrangements with organizations such as the Copyright Clearance Center and the Copyright Licensing Agency, can be found at our website: www.elsevier.com/permissions.

This book and the individual contributions contained in it are protected under copyright by the Publisher (other than as may be noted herein).

Notices
Knowledge and best practice in this field are constantly changing. As new research and experience broaden our understanding, changes in research methods, professional practices, or medical treatment may become necessary.

Practitioners and researchers must always rely on their own experience and knowledge in evaluating and using any information, methods, compounds, or experiments described herein. In using such information or methods they should be mindful of their own safety and the safety of others, including parties for whom they have a professional responsibility.

To the fullest extent of the law, neither the Publisher nor the authors, contributors, or editors, assume any liability for any injury and/or damage to persons or property as a matter of products liability, negligence or otherwise, or from any use or operation of any methods, products, instructions, or ideas contained in the material herein.

Library of Congress Cataloging-in-Publication Data
A catalog record for this book is available from the Library of Congress

British Library Cataloguing-in-Publication Data
A catalogue record for this book is available from the British Library

ISBN: 978-0-08-102633-5 (print)
ISBN: 978-0-08-102654-0 (online)

For information on all Woodhead publications
visit our website at https://www.elsevier.com/books-and-journals

Publisher: Matthew Deans
Acquisition Editor: Brian Guerin
Editorial Project Manager: Andrae Akeh
Production Project Manager: Joy Christel Neumarin Honest Thangiah
Cover Designer: Victoria Pearson

Typeset by SPi Global, India

Contents

Contributors		ix
Editor's Biography		xiii
1	**Introduction—Water**	**1**
	P. Senthil Kumar, P.R. Yaashikaa	
	1.1 Introduction	1
	1.2 Water and its sources	2
	1.3 Science of water	3
	1.4 Water consumption and use	5
	1.5 Water withdrawal	9
	1.6 Water pollution	9
	1.7 Water shortage	15
	1.8 Current challenges	17
	1.9 Conclusion	19
	References	19
2	**Water and Textiles**	**21**
	P. Senthil Kumar, K. Grace Pavithra	
	2.1 Introduction	21
	2.2 General process in dying industry	22
	2.3 Water consumption in textiles	25
	2.4 Water consumption in life cycle phases of textile products	30
	2.5 Fiber production phase	33
	2.6 Challenges and opportunities	35
	2.7 Conclusion	38
	References	38
3	**Water consumption in textile processing and sustainable approaches for its conservation**	**41**
	Kartick K. Samanta, Pintu Pandit, Pratick Samanta, Santanu Basak	
	3.1 Introduction	41
	3.2 Fiber, dye and process wise water requirement	43
	3.3 Strategies for reduction of water consumption in textile processing	45
	3.4 Textile processing using different irradiation techniques	53
	3.5 Effluent generation and treatment	54
	3.6 Conclusion and the future prospects	56
	References	57

4	**Water withdrawal and conservation—Global scenario**	**61**
	P. Senthil Kumar, C. Femina Carolin	
	4.1 Introduction	61
	4.2 Water withdrawal and conservation—a global scenario	63
	4.3 Global laws for water withdrawal	68
	4.4 Standards for water withdrawal and consumption	69
	4.5 Challenges and opportunities	70
	4.6 Conclusion	72
	References	73
	Further reading	75
5	**Water footprint management in the fashion supply chain: A review of emerging trends and research challenges**	**77**
	Eirini Aivazidou, Naoum Tsolakis	
	5.1 Introduction	77
	5.2 Literature background	79
	5.3 Trends and challenges for water footprint management in fashion supply chains	84
	5.4 Discussion and conclusions	87
	References	89
6	**Water footprint in fashion and luxury industry**	**95**
	Alice Brenot, Cécile Chuffart, Ivan Coste-Manière, Manon Deroche, Eva Godat, Laura Lemoine, Mukta Ramchandani, Eleonora Sette, Caroline Tornaire	
	6.1 Introduction	95
	6.2 The impact of the fashion industry on water resources	97
	6.3 Water awareness in luxury and fashion	99
	6.4 Solutions	106
	6.5 A favorable context	110
	6.6 Overview of African fashion and luxury: Showing the way!	110
	6.7 Conclusion and challenges	111
	References	112
	Further reading	112
7	**Analysis of water consumption and potential savings in a cotton textile dye house in Denizli, Turkey**	**115**
	Fatma Filiz Yıldırım, Barış Hasçelik, Şaban Yumru, Sema Palamutcu	
	7.1 Introduction	115
	7.2 Literature review	118
	7.3 Method	120
	7.4 General findings	127
	7.5 Conclusion	133
	References	133
	Further reading	134

8	**Water conservation in textile wet processing**	**135**
	M. Gopalakrishnan, V. Punitha, D. Saravanan	
	8.1 Introduction	135
	8.2 Low wet pickup techniques	135
	8.3 Foam finishing	140
	8.4 Cold pad batch method	142
	8.5 Dyeing machines	142
	8.6 Reduction of water in washing treatment	143
	8.7 Solvent dyeing	143
	8.8 Environmental effects and waste minimization	148
	8.9 Conclusion	150
	References	150
9	**Water requirement and sustainability of textile processing industries**	**155**
	A.S.M. Raja, A. Arputharaj, Sujata Saxena, P.G. Patil	
	9.1 Introduction	155
	9.2 Sustainability	156
	9.3 Water availability in India	157
	9.4 Water-conserving techniques for textile wet processing	161
	9.5 Waterless chemical processing of textiles	167
	9.6 Conclusion	170
	References	171
10	**Advances in the sustainable technologies for water conservation in textile industries**	**175**
	Luqman Jameel Rather, Salman Jameel, Ovas Ahmad Dar, Showkat Ali Ganie, Khursheed Ahmad Bhat, Faqeer Mohammad	
	10.1 Introduction	175
	10.2 Importance of water conservation in textiles	176
	10.3 Sustainable strategies for water conservation in textiles	180
	10.4 Challenges and opportunities for textiles and fashion	188
	10.5 Conclusion	189
	Acknowledgments	189
	References	189
	Further reading	194
Index		**195**

Contributors

Eirini Aivazidou Laboratory of Statistics and Quantitative Analysis Methods, Division of Industrial Management, Department of Mechanical Engineering, Aristotle University of Thessaloniki, Thessaloniki, Greece; Department of Computer Science and Engineering, School of Engineering and Architecture, Alma Mater Studiorum – University of Bologna, Bologna, Italy

A. Arputharaj Chemical and Biochemical Processing Division, ICAR-Central Institute for Research on Cotton Technology, Mumbai, India

Santanu Basak Chemical & Biochemical Processing Division, Central Institute for Research on Cotton Technology, Mumbai, India

Khursheed Ahmad Bhat CSIR Bio-Organic Division, Indian Institute of Integrative Medicine, Srinagar, India

Alice Brenot Luxury & Fashion Management, Skema Business School, Lille, France

Cécile Chuffart Luxury & Fashion Management, Skema Business School, Lille, France

Ivan Coste-Manière Luxury & Fashion Management, SKEMA Business School, Suzhou, China; Luxury & Fashion Management, SKEMA Business School, Sophia Antipolis, France; Luxury & Fashion Management, SKEMA Business School, Belo Horizonte, Brazil; Sil 'Innov & Eytelia, Courcelles, Belgium

Ovas Ahmad Dar Department of Chemistry, Jamia Millia Islamia (A Central University), New Delhi, India

Manon Deroche Luxury & Fashion Management, Skema Business School, Lille, France

C. Femina Carolin Department of Chemical Engineering, SSN College of Engineering, Chennai, India

Showkat Ali Ganie Department of Chemistry, Jamia Millia Islamia (A Central University), New Delhi, India

Eva Godat Luxury & Fashion Management, Skema Business School, Lille, France

M. Gopalakrishnan Department of Textile Technology, Bannari Amman Institute of Technology, Sathyamangalam, India

K. Grace Pavithra Department of Chemical Engineering, SSN College of Engineering, Chennai, India

Barış Hasçelik Pamukkale University, Engineering Faculty Textile Engineering Department, Denizli, Turkey

Salman Jameel CSIR Bio-Organic Division, Indian Institute of Integrative Medicine, Srinagar, India

Laura Lemoine Luxury & Fashion Management, Skema Business School, Lille, France

Faqeer Mohammad Department of Chemistry, Jamia Millia Islamia (A Central University), New Delhi, India

Sema Palamutcu Pamukkale University, Engineering Faculty Textile Engineering Department, Denizli, Turkey

Pintu Pandit Department of Fibres & Textile Processing Technology, Institute of Chemical Technology, Mumbai, India

P.G. Patil Chemical and Biochemical Processing Division, ICAR-Central Institute for Research on Cotton Technology, Mumbai, India

V. Punitha Department of Textile Technology, Bannari Amman Institute of Technology, Sathyamangalam, India

A.S.M. Raja Chemical and Biochemical Processing Division, ICAR-Central Institute for Research on Cotton Technology, Mumbai, India

Mukta Ramchandani United International Business School, Zurich, Switzerland; NEOMA Business School, Reims, France; Founder & CEO Moraltive, Zürich, Switzerland; Professor of Marketing at United International Business School, Zurich, Switzerland

Luqman Jameel Rather Department of Computer Science and Engineering, University of Kashmir, Baramullah; Department of Chemistry, Jamia Millia Islamia (A Central University), New Delhi, India

Kartick K. Samanta Mechanical Processing Division, National Institute of Research on Jute and Allied Fibre Technology, Kolkata, India

Pratick Samanta Department of Textile Technology, Indian Institute of Technology-Delhi (IIT-Delhi), New Delhi, India

D. Saravanan Department of Textile Technology, Bannari Amman Institute of Technology, Sathyamangalam, India

Sujata Saxena Chemical and Biochemical Processing Division, ICAR-Central Institute for Research on Cotton Technology, Mumbai, India

P. Senthil Kumar Department of Chemical Engineering, SSN College of Engineering, Chennai, India

Eleonora Sette Luxury & Fashion Management, Skema Business School, Lille, France

Caroline Tornaire Luxury & Fashion Management, Skema Business School, Lille, France

Naoum Tsolakis Centre for International Manufacturing, Institute for Manufacturing (IfM), Department of Engineering, School of Technology, University of Cambridge, Cambridge, United Kingdom

P.R. Yaashikaa Department of Chemical Engineering, SSN College of Engineering, Chennai, India

Fatma Filiz Yildirim Ozanteks Tekstil San ve Tic A.S R&D Center, Denizli, Turkey

Şaban Yumru Ozanteks Tekstil San ve Tic A.S R&D Center, Denizli, Turkey

Editor's Biography

Dr. Subramanian Senthilkannan Muthu holds a PhD in textiles sustainability and has written around 65 books and 80 research publications. He is well known for his contributions in the field and has extensive academic and industrial experience. He currently heads the department of sustainability for SgT and API and is based in Hong Kong. He earned his diploma, bachelor's, and master's in textile technology from premier institutes of India. He was awarded his doctorate from The Institute of Textiles and Clothing of The Hong Kong Polytechnic University. He has a decade of working experience in the arena of sustainability in textiles and clothing. He has worked with hundreds of factories in Asia and Europe on various aspects of sustainability. He was an outstanding student throughout his studies and earned numerous awards and medals, including many gold medals during his study. He is editor, editorial board member, and reviewer for many international peer-reviewed journals on textiles and environmental science disciplines. He is one of the directors of the Textile and Bioengineering Informatics Society (TBIS), a charitable organization created to foster, develop, and promote all aspects of science and technology in the bioengineering of materials, fibers, and textiles. He is the editor-in-chief of Springer's Textiles and Clothing Sustainability Journal.

Introduction—Water

P. Senthil Kumar, P.R. Yaashikaa
Department of Chemical Engineering, SSN College of Engineering, Chennai, India

1.1 Introduction

Water (H_2O) is the most valuable asset and an inexhaustible compound on the Earth's surface, covering in excess of 70% of the planet (Hossain, 2015). Water exists in three states in nature, namely, liquid, solid, and gas. It is in powerful balance between the fluid and gas states at standard temperature and pressure. At room temperature, it is tasteless, odorless, and colorless with a slight trace of blue. Numerous substances break down and dissolve in water and it is generally alluded to as the universal solvent. As a good solvent, water in nature may not be totally pure and its properties may shift marginally from those of pure water. In any case, there are likewise numerous compounds that are basically, if not totally, insoluble in water. Water is regularly found in each of the three basic states and it is necessary for all life on Earth. Water as a rule makes up 55%–78% of the human body (Hossain, 2015) and plays numerous critical roles therein. It is a major part of the vast majority of the body's cells, with the exception of fat cells, and it additionally cushions and lubricates the brain and the joints. It transports nutrients and diverts waste from the body cells. It additionally manages body temperature by redistributing heat from dynamic tissues to the skin and cooling the body through sweat. The greater part of the water in the body is found inside the cells of the body (around 66% is in the intracellular space), and the rest is found in the extracellular space, which is comprised of the spaces between cells and the blood plasma. Water is the principle constituent of the human body. It is typically around 60% of body weight in adult males, and is marginally lower, around 50%–55%, in females because of their higher muscle-to-fat ratio. The brain and muscles are around 75% water, the blood and the kidneys are around 81%, the liver is around 71%, the bones are around 22%, and fat tissue is around 20%. The body requires adequate water to survive and work properly (Hossain, 2015). People cannot live without drinking water for more than a couple of days—depending upon climate, movement levels, and different variables—while other nutrients might be disregarded for quite a long time, even months. No other nutrient is more basic or is required in such large quantities. Water is integral to the most fundamental physiological capacities, for example, directing pulse, body temperature, hydration, and digestion. A regular family unit utilizes a great deal of water. Many individuals use chemicals on their yards and gardens and afterward water them with unadulterated water. The water washes the chemicals off of the plants and after that may go down a storm drain straight to the waterways and streams in which fish make their homes. A great deal of water is required by farms. Water is used in hydroelectric plants, which use the dynamic energy of falling water to make power. Of all the power on the planet, around 20% is produced by hydropower (Hossain, 2015). Hydropower generation prevents a great deal of contamination. Hydropower generation is perfect and does not leave any waste. Hydropower reduces the amount of oil and coal required for power generation. Water is likewise fundamental in industry. It is heated and the steam is utilized to run hardware. Water is used to cool hot metal, for example, in the generation of steel. Water is likewise used to cool the air. It is a critical component in numerous items, such as chemicals, drugs, salves, shampoos, beautifying agents, cleaners, and also drinks. Water is utilized as a part of preparing food and in multitudinous plants and modern procedures, including paper manufacture. Water used in food and beverages must be completely pure, while different enterprises, for example, an assembly plant, may utilize a lower quality of water (Chaplin, 2001). This chapter includes a detailed discussion of water and its sources, measures of water consumption and utilization, methods involved in determining water usage, water withdrawal techniques, how groundwater and surface water are polluted, treatment techniques, and the challenges in controlling and maintaining water quality. Because water is an important resource, maintaining the quality of water remains as a major concern.

1.2 Water and its sources

Surface and ground water are the two noteworthy sources of water. Surface water consists of waterways, streams, lakes, and wetlands. Ground water is located in the pore spaces inside rocks and alluvium, in cracks, and in arrangement openings or courses in territories underlain by solvent carbonate rocks.

1.2.1 Surface water

Surface water begins for the most part as precipitation and is a blend of surface runoff and ground water. It incorporates large rivers, lakes, and the little upland streams that may start from springs and gather runoff from the watersheds. The amount of runoff depends on countless variables, the most vital of which are the amount and force of precipitation, the atmosphere and vegetation, and additionally, the land, geology, and the geographical highlights of the area. The nature of surface water is represented by its content of living things and by the measures of mineral and organic materials that it might have acquired. As rain falls through the environment, it cleans the air and assimilates oxygen and carbon dioxide. While streaming over the ground, surface water gathers sediment and particles of organic material, some of which will at last go into solution. It likewise grabs more carbon dioxide from the vegetation and smaller-scale living beings and microscopic organisms from the topsoil and from rotting matter. In inhabited watersheds, contamination may incorporate fecal material and pathogenic life forms, as well as other human and mechanical waste that has not been appropriately discarded. In provincial territories, water from little streams draining disconnected or uninhabited watersheds may have satisfactory bacteriological and compound quality for human use in its regular state. In many cases, surface water is liable to contamination and infiltration by pathogenic life forms and thus cannot be used by humans without treatment. It should be noted that reasonable water is not really fit for human use and that one can't depend completely on self-cleaning to create potable water.

1.2.1.1 Benefits of utilizing surface water as a noteworthy source of water

- Easily found and no complex hardware is required for finding a surface water source.
- Surface water is for the most part gentler than groundwater, which makes treatment considerably less complex.

1.2.1.2 Challenges

- Easily contaminated with organisms and chemicals that cause waterborne contamination and dangerous ailments.
- Turbidity frequently changes with the amount of precipitation. An increase in turbidity increases the treatment cost and operational cost.
- The temperature of surface water additionally changes with the temperature of the surroundings.

1.2.2 Groundwater

Groundwater is the water that fills subterranean pore spaces and fractures. Groundwater occurs in various land developments. Almost all stones in the upper portion of the Earth's outside layer, irrespective of their nature, have openings called pores or voids. Groundwater streams underground and can disintegrate shake, particularly limestone, forming sinkholes. Because groundwater is not visible, its nature and type are not well understood. Groundwater is easily tainted by human movement. Groundwater passes through water infiltration into the subsurface. Some dissipates, some is taken up by plants, some wets the surfaces of particles, and some permeates to the water table. Groundwater provides 66% of the world's freshwater supply.

1.2.2.1 Benefits of utilizing groundwater as a source of water

- Not as effectively contaminated or polluted as surface water.
- The nature of groundwater is consistently steady.
- Groundwater sources are mostly lower in bacteria than surface water sources.

1.3 Science of water

Water is a fluid substance made of atoms—a solitary, huge drop of water weighing 0.1 g contains around 3 billion trillion of them. Every molecule of water is comprised

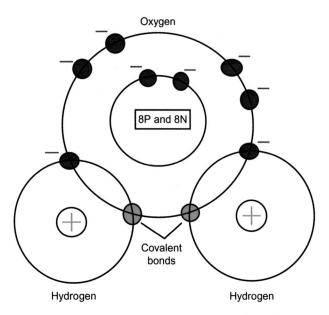

Fig. 1.1 Structure of water.

of three atoms: two hydrogen atoms in a kind of triangle with one oxygen atom—giving us the popular substance equation H_2O (Fig. 1.1). The marginally imbalanced structure of water atoms implies they draw in and stick to a wide range of substances. That is additionally why a wide range of substances will break down in water, which is now and then called a "universal solvent." Water can even disintegrate the strong rocks from which our planet is made, however, the procedure takes many years, decades, or even hundreds of years. The majority of the water in our reality is a mix of "customary" hydrogen molecules with "standard" oxygen particles, yet there are really three distinct isotopes of hydrogen and each of those can combine with oxygen to give an alternate sort of water. On the off chance that deuterium (hydrogen whose atoms contain one neutron and one proton rather than only one proton) joins with oxygen called heavy water, which is around 10% heavier than typical water. Also, tritium (hydrogen with two neutrons and one proton) can join with oxygen to make something many refer to as super heavy water. A unique aspect of water in our general surroundings is that it exists in three altogether different structures: solid, fluid, and gas. Standard, fluid water is the most well-known to us because water is a fluid under regular conditions, yet in addition we are familiar with solid water (ice) and vaporous water (steam and water vapor). Changing water between these three unique states is accomplished with surprising simplicity by changing its temperature or pressure (Sharp, 2001).

1.3.1 Characteristics of water

S.no	Factor	Characteristics	Measurement
1.	Physical	Turbidity	Defined by the quantity of suspended matter in mg/L of water and is measured using turbid meter
		Color	Usually colorless, depends on the use and is determined using tintometer
		Temperature	Depends on climate and it varies between 10°C and 25°C for potable water
		Taste and odor	Depends on the particulate matter present in the water
		Specific conductivity	Measured by the total amount of dissolved salts present in water
2.	Chemical	Total solids	Defined as the total amount of suspended and dissolved solids present in the water
		pH	Measured using potentiometer and if pH < 7, it is acidic and pH > 7, it is basic
		Hardness	May be temporary or permanent caused by chlorides, sulfates and nitrates of magnesium or calcium in water
		Chemical constituents	Presence of chlorides, nitrates, and heavy metals, such as lead, chromium, zinc, copper, etc.
		Dissolved gases and BOD	Oxygen is absorbed by water for consumption of organic matter present in water and determining the amount of oxygen resembles Biological Oxygen Demand (BOD)
3.	Biological	Microbes	Microscopic pathogens, such as bacteria, fungi, viruses, protozoa, and worms grow in water, causing infection

1.4 Water consumption and use

A huge difference exists between the two terms "water consumption" and "water use." Water use concerns the aggregate sum of water extracted from its source to be utilized. The quantity of water use helps us to assess the level of need from various fields, such as agriculture, industry, and for domestic purposes, while water consumption is the part of water use that does not return to the first water source once taken out. Utilization occurs when water is lost into the air through dissipation or absorbed into a plant and thus not available for reuse. Water utilization is especially significant for determining water shortages. Offstream water use occurs when water is withdrawn from a surface or groundwater source and transported to the place of use. After use, the wastewater is passed to either a wastewater treatment office or is returned to the hydrologic framework. Instream use occurs when the water stays in streams (surface water) and aquifers (groundwater) during use, for example, in hydroelectric power production. In spite of the fact that there is significant variety in how water is utilized, there are fundamental similarities for deciding the volume or rate of utilization. An initial phase in deciding water utilization is to characterize the approach needed to meet the requirements. This first, or arranging stage, considers the required level of detail, the time and labor that is available, the clients and their attributes, and the availability of water utilization information from different sources. Once the approach has been resolved, techniques can be chosen to gather the information in the most exact and effective way possible.

In the previous two decades, open water providers, ventures, and irrigators have been more engaged with water protection as a method for either extending administrations with a limited supply or addressing present demands with a diminishing supply. Withdrawals by some open water providers have diminished through the use of (1) spill discovery programs, (2) residential preservation programs, including state-funded training, (3) protection counseling administrations to modern and business clients, and (4) expanded rates for water supply and wastewater treatment. Industrial operations have reduced water withdrawals by overhauling their plant tasks to increase the amount of water reused and utilizing new advanced techniques that require less water. Numerous irrigators are using low-weight dribble and stream frameworks rather than high-weight splash frameworks. Zones with huge occasional climatic variety may encounter lessened regular variety through confinements or protection programs that influence use. Temporary limitation programs during dry season can diminish water use uniquely; when restrictions are not changeless, people tend to relapse to past water-use patterns. A few groups have unchanging water use limitations, particularly in bone-dry states. Preservation programs, including the retrofit of older structures with low-stream systems, supplanting grass with low-water-use finishing, and expanding costs for metered water clients, have a more permanent effect on water use designs. Water likewise can be reused as wastewater recovery after discharge from a wastewater treatment plant. The diminishing accessibility of freshwater and the increased treatment costs encourage the adoption of wastewater recovery for water systems or even for business and mechanical employments (Marcus, 2012).

1.4.1 Methods for determining water use

Water quality and water consumption are important factors to be monitored and measured so that water demand can be controlled. The following are steps involved in determining water use (Juneja and Chauhdary, 2013):

- Primary data collection;
- Secondary data collection;
- Data analysis.

1.4.1.1 Primary data collection

Essential information can be acquired by coordinate strategies, for example, utilizing total water meters, or by (a) estimating the drawing rate with quick stream meters, and (b) estimating or evaluating pumping length. In coordinate techniques, a cumulative meter records the volume of water passing through the meter in a similar fashion to an odometer. The meter readings increment until the highest number on the meter is reached, whereupon the meter dial returns to zero and starts once more. These meters are regularly utilized by open water providers to record the quantity of gallons or cubic feet the user has gotten since the meter was last checked. The precision and reliability of these meters fluctuate quite a bit. While depending on these meters for data, the accompanying elements should be considered: (1) volume for which the meter was designed; (2) precision of the meter; (3) nature of the meter; (4) the date of the latest adjustment; and (5) whether the meter will overestimate or underestimate as the meter wears. Indirect strategies for estimating water consumption make use of prompt stream meters to gauge the volume of water currently going through a pipe. These meters can be forever situated or might be convenient. Quick meter readings should be combined with a running-time estimation, keeping in mind the end goal of determining water use over a day, month, developing season, or year. Sonic stream meters might be the quickest and easiest strategy for deciding quick pipe stream. Compact sonic stream meters can be attached to the outside of a pipe and adjusted, record the stream rate, and then be removed in typically less than an hour with no obstruction to the stream or the pumping task. A few different types and brands are available from different producers. Stream rates may change at the time of pumping depending upon the drawdown in the well and possibly framework alterations; thus, a few estimates may be expected to determine a normal stream rate after some time.

1.4.1.2 Secondary data collection

Auxiliary information procurement includes accumulation assessment and investigation of estimated or evaluated information sent by water clients to state and federal offices. The information might be routinely sent to a state or federal organization through legal necessity or approach prerequisite. The information is ordinarily gathered in light of a particular requirement for the information by a federal or state office and reaction to the structures might be deliberate. In the underlying phases of water use information gathering, there may be a check with other federal and state offices to decide the accessibility of the announced information and finished or arranged reviews. Detailed and studied water use information must be utilized carefully in light of the fact that the precision and quality of individual withdrawal or release reports fluctuates quite a bit. Revealed or overviewed water use information might be either metered or evaluated. Metered information, for the most part, is dependable. The quality of the information may be less precise if the plan stream of the meter is not inside the actual stream of the meter, if the meter has not been aligned recently, or in the event that it is inexpertly read. A few clients report the real reading from a total meter, which would show an expanded use every month. Errors may occur when transferring readings from the meter to the field sheet and to the information base. Detailed or overviewed water use information that depends on gauges requires an assessment of the estimation technique. Now and again, the revealed water use may adjust all the more closely with withdrawal and release allows instead of to real utilization. Appraisals of aggregate water use rely upon precise reports, and a satisfactory number of returned surveys. Revealed or reviewed water use information is occasionally confirmed in the field (Weerakkody et al., 2017).

1.4.1.3 Data analysis and management

The information determined relies upon a coefficient to evaluate water use or an estimate figured from models, such as those including extrapolation, nonfinancial numerous coefficient models, and econometric models. The coefficients might be determined from writing, refined using optional information, or created from essential information gathering. Models that use extrapolation methods in particular are proper for evaluating water use in circumstances in which extraordinary precision and detail are not required. Extrapolating water use values through time depends on the assumption that water use can be found by the water-use slopes used before. Time is the free factor and water use is the dependent variable. Models in light of extrapolation incorporate basic relapse and pattern investigation. Noneconometric models consider water use as a numerical variable of at least two logical factors, such as climate or socioeconomics, but exclude the cost or the financial elements. Water use is viewed as a vital prerequisite unaffected by cost. The factors are fused into a model that fits recorded information and the coefficients are evaluated, as a rule, by regression. Econometric models depend on financial elements and the interest for water is evaluated as a component of the cost of water and other monetary variables. Multivariate models determine water use need as an element of climate, financial aspects, and socioeconomics, and the capacity is measurably decided. Econometric models are particularly suitable for assessing complete open water supply withdrawals and conveyances (Ahmed et al., 2013). Water consumption and use can be depicted as follows.

Withdrawals + open water supply conveyances = water use

Water use − immoderate use = wastewater accumulation + return stream

Data administration is a basic part of water use information accumulation, assemblage, investigation, and utilize. There are a few distinct categories of water use information: (1) distinguishing proof, (2) geographic, (3) hydrologic, and (4) rate or volume. Recognizable proof information incorporates the name, address, and distinguishing proof (ID) numbers that tie together extraordinary informational collections by utilizing the allowance numbers or other exceptional numbers appointed to clients by various information gatherers. Geographic data is basic when a Geographic Information System (GIS) is utilized. A scope longitude locates the client or other purpose of interest and recognizes the district, state, and watershed in which the user is found. Hydrologic data recognizes the assets influenced, for example, the waterway, watershed, or aquifer. Withdrawals from every key aquifer in a zone are valuable data. Information on withdrawal destinations, for example, drillers' well logs, might be accessible from the USGS ground-water documents in state district workplaces. Data likewise might be accessible from state contamination control organizations or state general wellbeing offices. Water use information can be gathered in a few unique units (Dawes et al., 2016).

1.5 Water withdrawal

Water withdrawals, or water reflections, are characterized as freshwater taken from ground or surface water sources, either for all time or incidentally, and transported to the location of use. On the off chance that the water is returned to a surface water source, reflection of a similar amount of water by the downstream client is included again and this may cause double counting. The information incorporates deliberations for open water supply, water system, mechanical procedures, and cooling of electric power plants. Mine water and seepage water are incorporated, while water utilized for hydroelectricity is ordinarily rejected. Water withdrawal indicates the aggregate sum of water withdrawn from a surface water or groundwater source. Estimates of the amount of water withdrawn help assess requests from residential, modern, and horticultural clients. Estimates of water withdrawal show the level of rivalry and reliance on water assets. Withdrawing water faster than it can be replenished can may mean an inability to meet both present and future needs. Abuse of water can likewise hurt environments, especially those that rely upon shallow ground water or lasting streams. Requests that surpass supply can prompt clashes between clients, among upstream and downstream groups, and between withdrawal needs and recreational and natural interests (Hossain, 2015).

Advantages of withdrawing groundwater
- Helpful for drinking and water system
- Effortlessly accessible and exists all over the place
- Inexhaustible if not overpumped or contaminated
- No evaporation loss
- Less expensive to remove than most surface water

Impediments
- Aquifer exhaustion and contamination from overpumping for a long time
- Sinking of land from overpumping
- Saltwater intrusion into drinking water supplies close to seaside territories
- Reduced water streams to surface waters
- Increased cost and contamination from deeper wells

1.6 Water pollution

Water contamination may be characterized as the tainting of streams, lakes, oceans, underground water, or seas by substances that are harmful to living creatures. Industrialization and overpopulation are two necessary components for water contamination. Water contamination occurs when undesirable materials go into the water, changing its nature and making it unsafe for human wellbeing. Water is an essential asset utilized for drinking and other basic purposes in our lives. Safe drinking water is vital for human wellbeing everywhere throughout the world. Being an all-inclusive solvent, water is a noteworthy wellspring of disease. As noted by the World Health Organization (WHO) 80% of infections are waterborne. Water consumption in different nations does not meet WHO principles. 3.1% of deaths occur due to the unhygienic and low quality of water. The release of residential and modern waste spillage from water tanks, marine dumping, radioactive waste, and climactic effects are significant reasons for water contamination. Heavy metals from runoff and mechanical waste can amass in lakes and streams, making them unsafe to people and creatures (Bibi et al., 2016). Poisons in modern waste are the real reason for insusceptible concealment, regenerative disappointment, and intense harm. Various illnesses, such as cholera, typhoid fever, gastroenteritis, loose bowels, regurgitating, and skin and kidney issues, are spreading through contaminated water. Human wellbeing is influenced by the immediate harm of plant and creature nourishment. Water toxins are killing ocean weeds, mollusks, marine birds, fish, scavengers, and other ocean life forms that serve as food for humans. Bug sprays, such as DDT, have moved up the food chain. These bug sprays are destructive for people. It is estimated that 75%–80% of water contamination is caused by residential sewage. Waste from the businesses, such as, sugar, materials, electroplating, pesticides, mash, and paper also contaminate water. Polluted waterways may have an unpleasant odor and contain less greenery. Eighty percent of the total populace is confronting dangers to water security. A great deal of residential sewage is discharged into waterways and a large portion of the sewage is untreated. Household sewage contains toxins, strong waste, plastic litter, and bacterial contaminants and these dangerous materials cause water contamination. Diverse modern effluent that is discharged into a stream without treatment is a significant reason for water contamination. Risky material released from businesses is a source of surface water and groundwater pollution. The type of contaminant depends on the nature of the business. Lethal metals enter in to water and decreased the nature of water. Twenty five percent of contamination is caused by businesses and is more harmful. An expanding population is causing numerous issues and one of them is that it plays a negative part in water contamination. An expanding population causes an increase in strong waste. Strong and fluid waste is released into streams. Water is additionally sullied by human excreta. In contaminated water, a substantial number of microscopic organisms are likewise present, which is destructive for human wellbeing (Halder and Islam, 2015). Government is inadequate to supply basic needs to citizens on account of the increasing population. Sanitation offices are located more in urban regions than in country zones. Polythene packs and plastic waste are a noteworthy source of contamination. Waste is discarded by placing it into plastic bags. It is assessed that 3% of individuals in urban territories do not have access to sanitary facilities. Seventy seven percent of individuals use flush lavatories and 8% use pit toilets. Urbanization can cause numerous ailments. Congestion, unhygienic conditions, and perilous drinking water are real medical problems in urban zones. One fourth of the urban populace is susceptible to sickness. Pesticides are utilized to eliminate microorganisms, bother and diverse germs. Substances containing pesticides are dirtying the water and influencing the nature of water. In the event that pesticides are abundant or used ineffectively then it would be unsafe for the agribusiness biological community. Just 60% of manures are utilized as a part of the dirt different chemicals filtered in to soils dirtying the water; cyanobacteria are rich in dirtied water and abundance phosphate keep running off prompts eutrophication. Chemicals build up and blend with waterway water because of flooding, substantial precipitation, and overabundance; once in the water system, they may enter the food chain. These chemicals are deadly to life forms and numerous vegetables and natural products are contaminated with these chemicals. Pharmaceuticals in water additionally cause water contamination and this is perilous to human wellbeing (Owa, 2013).

1.6.1 Consequences of water pollution

There is a noteworthy relationship between contamination and medical issues. Microorganisms that cause sickness are known as pathogens and these pathogens spread sickness among people. A few pathogens are found everywhere, while some are found in specific zones. Numerous waterborne maladies spread between people. Substantial precipitation and floods are identified with outrageous climate and causing distinctive maladies for developed and developing nations. Ten percent of the populace relies upon nourishment and vegetables that are grown in contaminated water. Numerous waterborne sicknesses are connected with fecal contamination of water sources and this results in fecal-oral course of disease (Desai and Vanitaben, 2014). The health hazards related to contaminated water incorporate diverse ailments, for example, respiratory ailments, tumors, diarrheal maladies, neurological confusion, and cardiovascular infection. Nitrogenous chemicals result in growth and blue child disorder. The death rate because of growth is higher in rural zones than urban regions because urban residents utilize treated water for drinking while provincial individuals do not have access to treated water and thus utilize natural water. Destitute individuals are at more serious danger of infection because of inappropriate sanitation, cleanliness, and water supply. Contaminated water has large negative impacts on women who are exposed to chemicals during pregnancy; it causes an increased incidence of low birth weight and accordingly fetal wellbeing is influenced. Low-quality water obliterates the yield generation and contaminates our food, which is perilous for sea life and human life. Toxins bother the evolved way of life and overwhelming metals, particularly press influences the respiratory arrangement of fish. Iron accumulation in fish gills deadly to them and when these fish are eaten by humans, significant medical problems can result. Metal-debased water prompts male pattern baldness, liver cirrhosis, renal disappointment, and neural issues (Kamble, 2014).

1.6.2 Factors determining contamination in water

S.no	Factors	Nature
1.	Physical	Color, odor, turbidity, taste, temperature, and electrical conductivity
2.	Chemical	Carbonates, sulfates, chlorides, fluorides, nitrates, and metal ions that constitute the total dissolved solids
3.	Biological	Bacteria, fungi, algae, protozoa and virus

1.6.3 Sources of water pollution

Water toxins are those substances that make any physical, compositional or natural change in the water body. They have negative impacts on living creatures. Water contamination might be surface or groundwater contamination. The real sources of water contamination may be point, nonpoint, characteristic and anthropogenic sources. Identifiable sources that discharge toxins or effluents specifically into various water bodies of new water are called point sources. Household and mechanical waste are examples of this. The point sources of contamination can be successfully checked. Then again, the nonpoint sources of water contamination are scattered or spread over large regions. These kinds of sources convey poisons in a roundabout way through ecological changes and are responsible for the greater part of the contaminants in streams and lakes. An expansion in the amount of naturally occurring substances is named characteristic contamination. Human activity that results in the contamination of water is called an anthropogenic or artificial source of water contamination. In the greater part of our towns and numerous townships, groundwater is the main source of drinking water. Thus, contamination of groundwater is a genuine concern. Groundwater gets contaminated in various ways. The permeable layers of soil keep down strong particles while the fluid is permitted to go through. Dissolvable poisons can blend with the groundwater. Notwithstanding these, the frequent use of nitrogenous composts and the unchecked arrival of harmful waste, and even cancer-causing substances, by modern equipment may infiltrate through the surface and blend with groundwater. This issue is intense, particularly in regions in which water is accessible close to the surface of the earth. Because the development of groundwater through the permeable shake is moderate, contaminations which get blended with the groundwater are not promptly weakened. Moreover, groundwater does not approach air (as compared to surface water) and thus oxidation of toxins into innocuous substances in groundwater does not happen (Kamble, 2014).

1.6.4 Biological effects of water pollution

1.6.4.1 Eutrophication

Eutrophication is the process of excessive nutrient enrichment of waters that typically results in problems associated with macrophyte, algal or cyanobacterial growth.

1.6.4.2 Biochemical oxygen demand

The amount of oxygen spent by microorganisms at 27°C and in haziness for 3 days in separating natural waste in a water body is called its organic oxygen request or Biochemical Oxygen Demand (BOD). The microorganisms exhibit in the framework follow up on this loss for their own particular utilization and development. In the process the metabolic movement requires oxygen which is met by the broken-down oxygen show in water. It is this measure of oxygen which is characterized as organic oxygen request BOD. The BOD estimation of an amphibian framework relies on:

- the sort and measure of natural waste
- the living beings following up on it
- temperature and pH

The more natural waste in the body of water, the larger is the quantity of oxygen required to separate it organically and thus the BOD estimation for the water is higher. This value is a decent measure in assessing the level of contamination in a water body. Less polluted water has a lower BOD. Its estimate is utilized as a model for overseeing water contamination of a water body. An assessment is made by measuring oxygen demand in water when standing still at 20°C for 5 days.

1.6.4.3 Biomagnification

An assortment of lethal chemicals may spread through the food chain. Dangerous pesticides might be sprayed for controlling bugs and parasites; however, they also do damage to other (nontarget) living beings. For instance, DDT was used in the United States to control mosquitoes in an amount that was anticipated to be safe to nontarget creatures, such as fish and birds. DDT aggregated in swamps and microscopic fish. These tiny fish were eaten by larger fish and the fish then had a higher concentration of DDT in its body. Further, when birds ate the fish, they collected still higher amounts. This increasing concentration of aggregated lethal chemicals as one goes higher up the food chain is called biomagnification. Biomagnification has, on occasion, undermined the proliferation and survival of carnivores at the top of the food chain.

1.6.5 Control and treatment

The intent of contamination control is to control the discharges of effluents into the air, water, and land. Without contamination control, the waste items from used products, warming, farming, mining, assembling, transportation, and other human exercises, regardless of whether they gather or scatter, will debase nature. Contamination avoidance and waste minimization are more attractive than contamination control. In any case, contamination may be limited by following these practices: (1) by reusing, (2) waste minimization, (3) by moderating, (4) by anticipating, and (5) by compost. Direction and observing is a viable method for contamination administration. Numerous countries worldwide have passed laws to manage different kinds of contamination and additionally to relieve the negative impacts of contamination.

Wastewater produced by the actions of a family unit, business, or landfill is called sewage, which is called metropolitan water contamination. Sewage contains strong issues, such as suspended colloidal and broken up natural issue, cleanser, mineral issue, supplements, and gases. Sewage is one of the real reasons for waterborne infections and along these lines the treatment of sewage is one of the critical undertakings. For quite a while, treatment of metropolitan waste as sewage included predominantly the removal of suspended solids, oxygen consuming materials, and harmful microscopic organisms. Presently the transfer of the strong buildup from sewage has been enhanced by applying civil treatment forms (Alrumman et al., 2016). The treatment of this wastewater is done in the following three phases:

- (i) Primary treatment
- (ii) Secondary treatment
- (iii) Tertiary treatment

In a treatment plant, the waste is passed through a progression of screens, chambers, and substance procedures to reduce its mass and poisonous quality. During this essential treatment, a large amount of suspended solids and inorganic material is removed from the sewage. The auxiliary stage reduces the amount of natural material by speeding up common organic procedures. Tertiary treatment is done when water is to be reused. Here 99% of solids are removed and different substance forms are utilized to guarantee that the water is free from contaminating materials.

1.6.5.1 Primary treatment

At the point at which the wastewater is to be dumped off into a waterway or steam, the treatment is completed by sedimentation, coagulation, and filtration. This is known as an essential treatment. In the event that the water is required for drinking purposes, it needs to experience a further treatment called auxiliary and tertiary medications.

1.6.5.2 Secondary treatment

After essential treatment, the water still is not fit for drinking purposes and requires assist treatment. This is done through auxiliary or natural treatment. A general strategy is to enable dirtied water to spread over an extensive bed of stones and rock so the development of various microorganisms requiring supplements and oxygen is supported. Over some undefined timeframe, a quick moving natural pecking order is setup. For instance, microorganisms expend natural issue from the dirtied water; protozoa live on microscopic organisms. Different life forms, including green growth and parasites, help in the cleaning process. This is called auxiliary treatment of water. The procedures engaged with auxiliary treatment are softening and air circulation.

1.6.5.3 Tertiary treatment

Tertiary treatment really means sterilizing the water. Chlorine is the most regularly used disinfectant for eliminating microorganisms. In any case, chlorine additionally may react with residual material in the water to produce chlorinated hydrocarbons (poisonous and possibly cancer-causing). It is thus desirable to decrease the organic material in water before passing chlorine gas. Different strategies for purification, for example, UV radiation, ozone gas treatment, or reverse osmosis are favored over chlorine treatment. However, these strategies are costlier.

1.7 Water shortage

In spite of water being an existential requirement for people, it is additionally a standout among the most under organized yet finished manhandled products. Water is fundamental to our lives; however, it has not been the main issue at the center of our arrangements while we quickly develop into an urban culture. Early social orders comprehended the significance and requirement for water and arranged their lives around it. Human advancements were conceived and lost by virtue of water. Water shortages are for the most part manmade because of overabundant population development and depletion of water assets. The main reasons behind water shortages are:

- Wasteful utilization of water for farming.
- Reduction of conventional water energizing territories.
- Sewage and wastewater discharge into customary water bodies.
- The arrival of chemicals and effluents into waterways, streams, and lakes.
- Absence of effective water administration and conveyance of water between urban shoppers, the agricultural sector, and industry.

1.7.1 Causes of water shortages

1.7.1.1 Overconsumption of water

Water overuse is a huge issue that many individuals are managing. It might be abused on individuals, creatures, arrive, or any number of things. Water might likewise be used for recreation with no thought as to the impact that it might have on the general surroundings.

1.7.1.2 Water pollution

Water contamination is a serious issue, particularly in places that do not have proper drainage or sewage systems. Contamination can be anything from oil to corpses to chemicals to fecal matter. Regardless of what it is, it creates a considerable number of issues for the general population who need it.

1.7.1.3 Dispute

In the event that there is strife over a piece of land, it might be hard to get to the water that is located there. In the worst situations, individuals could wind up their lives in the event that they endeavor to get to the water in their locations. This can bring about an assortment of different issues, including contamination.

1.7.1.4 Drought

A drought occurs when a place does not receive enough rainfall to have the capacity to manage the life that is living there. A few regions are in an endless drought, while different places might manage a dry season now and again. Dry seasons are basically everywhere throughout the world, and there is little that can be done to keep such things from happening.

1.7.2 Water shortage in Asia

Financial and population development and environmental change could prompt genuine water shortages over a wide region of Asia by the year 2050, a recently distributed investigation by MIT researchers has found. Asia's water troubles are increasing. As of now the world's driest landmass in per capita terms, the mainland, now faces an extreme drought that has dried an immense area stretching from southern Vietnam to central India. The present dry season in parts of Southeast and South Asia is the worst in decades. Drought conditions may not bring down structures, but rather they convey high social and monetary expense. A huge number of Asians now face serious water deficiencies, and some have been compelled to move. In the interim, the mounting drought-related misfortunes in a portion of the world's best rice-growing nations—Thailand, Vietnam, and India—have undermined the world's now tight rice market. Asian nations must cooperate to guarantee more productivity in water utilization, increase the utilization of reused and desalinated water, and advance inventive arrangements that propel preservation and adjustment endeavors.

1.7.3 Water shortage solutions

1.7.3.1 Water recycling

There are many innovations that enable one to reuse rainwater and other water that might be used in the home. Consider finding out about how one can reuse water. This saves money and avoids the problems caused by water scarcity or shortage.

1.7.3.2 Awareness

There are a lot of chances out there for individuals to take in more of their general surroundings with a specific end goal. By teaching individuals who are not managing a water shortage, they can be in a situation to help. The individuals who are managing it can be taught how to keep the issue from becoming worse later on.

1.7.3.3 Recent techniques for water conservation

There has been a great deal of work in the area of water conservation, yet there is likewise a considerable amount that should be done to guarantee that whatever is left of the world can monitor water.

1.7.3.4 Improvement in sewage system

Clean drinking water begins with a good sewage framework. Without appropriate sanitation, the water in a region moves toward becoming ridden with sickness and any number of different issues. By enhancing the sewage frameworks in these regions, we can keep a water shortage from turning into something worse.

1.8 Current challenges

Water is progressively turning into a need strategy issue at the worldwide level. Expanding globalization is propelling the use of new guidelines and strategies for the universal exchange of products and ventures, mirroring the expanding impact of multinational firms connected with by implication in water utilize and exchanges. This globalization of exchange has boundless ramifications for customers, governments, and nature. While mass water isn't generally exchanged, with the exception of modest amounts in bottles, the water used to create the products that are exchanged, called virtual water, can affect water adjustments in a district in a major way. The effect of globalization on water may be considered from two different points of view: the negative consequences for water of the developing coordination of the world economy, specifically concerning water pollution and related ecological debasement and water itself as a protest of worldwide exchange approaches. Inflowing water quality is as critical as the amount of water. Environmental changes might be caused by minor water quality changes. Numerous contaminants regularly consolidate synergistically to cause enhanced, or unique, impacts compared with the combined impacts of toxins considered independently. Surges and dry spells can substantially affect the environments of wetlands and timberlands. Cycles of dry spells and surges are a characteristic part of biological communities; they change in accordance with and are affected by them. Surges and their related residue can revive regular biological communities, giving more plentiful water and rich soil for plants. Urbanization and other land utilization changes, poor rural practices, and industrialization are among those things that can change the amount of water and quality administrations in environments, and consequently negatively impact biological communities. Reducing the amount of needed water can likewise be accomplished by controlling different items that are not straightforwardly identified with water, but rather which are similarly vital. Water administrators are more significantly tested when managing transboundary water bodies, particularly universal ones. Transboundary rivers and aquifers are regular highlights of the present hydrologic and political scene. Inside various nations and gatherings of nations that offer an aquifer or waterway it is extremely valuable to find an agreement on how to share the water in the midst of worry before that pressure happens as opposed to work out such understandings amid times of pressure (Oki et al., 2006).

The water emergency in the 21st century is significantly more identified with administration than with a genuine emergency of shortage and stress. It is the consequence of an arrangement of ecological issues disturbed by monetary and social improvement issues. A solidified information base changed into an administration instrument can be a standout among the best types of defying the issue of water shortage, water pressure and disintegrating quality. The significant reasons for the water emergency are listed below:

- Intense urbanization, expanding the interest for water, growing the release of debased water assets, while there is enormous interest for water for drinking and monetary and social improvement.
- Water stress and shortage in numerous locales of the planet because of modifications in accessibility and increased requirements.
- Poor framework or foundation in a basic state in numerous urban zones, with failures in the system of up to 30% after treatment.
- Problems of stress and shortage because of worldwide changes with unusual hydrological events expanding the human populace's powerlessness and threatening food security (serious downpours and extraordinary periods of dry season).
- Problems caused by the absence of verbalization and absence of steady activities of manageability of water assets and for natural maintainability.

Notwithstanding the issue of administration of water assets, another component of administration ought to likewise experience noteworthy change: from the local, sectoral and reaction administration there is presently a progress to administration at the level of the biological community (watershed) that is coordinated (incorporating the cycle of air, shallow, and underground waters and incorporating the products employments). The interest of clients, people in general, the private part and the general population division ought to be one of the chief focal points of this administration of water assets with regards to watersheds. This investment ought to enhance and extend the maintainability of the free market activity and of the aggregate security of the populace in connection to accessibility and helplessness.

With developing urbanization and industrialization, the request of water for city and modern use has likewise been expanding. Water is a cross-cutting issue that pervades the social, financial, and political texture of India. India has exceedingly various atmospheric conditions by locale that can cause climate-related catastrophes, for example, serious surges and dry seasons. The nation's unpredictable climate patterns have a tendency to make hydrological surprises while water reaping, a typical source of water, ends up testing. In excess of 60% of India's flooded horticulture and 85% of drinking water supplies are subject to ground water. Notwithstanding, it is predictable that India's streams and other surface water sources are not ready to adapt to the developing interest for water the same number of such water sources are either dirtied or dried because of delayed dry spell. The present water framework in India has restricted the ability to manage new financial improvements. The general water industry is additionally challenged by the impact of environmental change and ecological contamination. Over the previous decade, noteworthy advances have been made to give groups access to water supplies. Be that as it may, numerous water systems are not working proficiently because of poor upkeep, older systems with no spending plan for retrofit and substitution, and pollution or exhaustion of water sources (Palaniappan et al., 2010).

1.9 Conclusion

Water is basic to people and other living things despite the fact that it gives no calories or natural supplements. Water is likewise fundamental to the mechanics of the human body. The body can't work without it, just as an automobile can't keep running without gas and oil. All of the cells and organs that make up our whole life systems and physiology rely upon water for their function. Water contamination is a worldwide issue and the world is confronting the awful consequences of dirtied water. Significant sources of water contamination are release of household and horticulture waste, population development, exorbitant use of pesticides and manures, and urbanization. Bacterial, viral, and parasitic infections are spreading through contaminated water and influencing human wellbeing. It is suggested that there ought to be appropriate waste transfer frameworks and waste ought to be dealt with before entering into a waterway. Instructive and awareness projects ought to be designed to control the contamination. Coordinated, forward-thinking administration with options for and change of the various uses must be embedded at the level of hydrographic bowls keeping in mind the end goal to decentralize administration and to enable cooperation among users and general society and private divisions. Training of all levels of the group and the readiness of directors with new methodologies is another vital improvement for water assets administration in the 21st century. Along these lines, water is the foundation of the immense biodiversity on our planet. Without water in adequate amounts and quality, the fate of both people and natural life is undermined. Water plays a role in every key procedure on Earth and is a foundation for life. Water is in this way the biggest fortune humanity has acquired from nature. New worthwhile procedures should utilize the water more proficiently in inundated horticulture, which will permit expanding food production without expanding water utilization to meet the issue of water shortages.

References

Ahmed, T., Scholz, F., Al-Faraj, W., et al., 2013. Water-related impacts of climate change on agriculture and subsequently on public health: a review for generalists with particular reference to Pakistan. Int. J. Environ. Res. Public Health 13, 1–16.

Alrumman, S.A., El-kott, A.F., Kehsk, M.A., 2016. Water pollution: source and treatment. Am. J. Environ. Eng. 6 (3), 88–98.

Bibi, S., Khan, R.L., Nazir, R., et al., 2016. Heavy metals in drinking water of Lakki Marwat District, KPK, Pakistan. World Appl. Sci. J. 34 (1), 15–19.

Chaplin, M.F., 2001. Water: its importance to life. Biochem. Mol. Biol. Educ. 29, 54–59.

Dawes, S.S., Vidiasova, L., Parkhimovich, O., 2016. Planning and designing open government data programs: an ecosystem approach. Gov. Inf. Q. 33 (1), 15–27.

Desai, N., Vanitaben, S., 2014. A study on the water pollution based on the environmental problem. Indian J. Res. 3 (12), 95–96.

Halder, J.N., Islam, M.N., 2015. Water pollution and its impact on the human health. J. Environ. Pollut. 2 (1), 36–46.

Hossain, M.Z., 2015. Water: the most important precious source of our life. Glob. J. Adv. Res. 2 (9), 1436–1445.

Juneja, T., Chauhdary, A., 2013. Assessment of water quality and its effect on the health of residents of Jhunjhunu district, Rajasthan: a cross sectional study. J. Public Health Epidemiol. 5 (4), 186–191.

Kamble, S.M., 2014. Water pollution and public health issues in Kolhapur city in Maharashtra. Int. J. Sci. Res. Publ. 4 (1), 1–6.

Marcus, Y., 2012. Supercritical Water. A Green Solvent: Properties and Uses. Wiley, Hoboken, NJ.

Oki, T., Valeo, C., Heal, K. (Eds.), 2006. Hydrology 2020: An Integrating Science to Meet World Water Challenges. International Association of Hydrological Sciences, Wallingford, UK.

Owa, F.D., 2013. Water pollution: sources, effects, control and management. Mediterr. J. Soc Sci. 4 (8), 65–68 (ISSN 2039-2117).

Palaniappan, M., Gleick, P.H., Allen, L., Cohen, M.J., Christian-Smith, J., Smith, C., 2010. Ross, N. (Ed.), Clearing the waters: a focus on water quality solutions. UNEP, Nairobi, Kenya. 89.

Sharp, K.A., 2001. Water: Structure and Properties. Encyclopaedia of Life SciencesJohn Wiley & Sons, Chichester.

Weerakkody, V., et al., 2017. Open data and its usability: an empirical view from the Citizen's perspective. Inf. Syst. Front. 19 (2), 285–300.

Water and Textiles

P. Senthil Kumar, K. Grace Pavithra
Department of Chemical Engineering, SSN College of Engineering, Chennai, India

2.1 Introduction

The textile industry is considered to be one of the most industrialized sectors in the manufacturing industry and the end uses are seen in clothing, home furnishings, and in industrial uses. Several subsectors, such a raw material production, semi-processed products, and final products, are involved in the textile industry. It is considered to be a diverse industry, which utilizes the agricultural sector for natural fibers, the chemical industry for manmade fibers, along with other industrial sectors, including optical fibers, air bags, insulation, roofing materials, and filters. Water plays a major role in textile industries. A large amount of water is used in all sectors of the industry. Almost all dyes and chemicals are applied to fabrics in the form of water baths. The various steps, such as desizing, scouring, bleaching, and mercerizing, use water in large quantities. The water thus used returns to our ecosystem without treatment. The dyes and chemicals used during the milling process are returned to our water streams and pollute the surface water as well as the ground water. A decrease in the amount of useable water occurs over time due to an increase in pollution. The people and the living beings in the ecosystem are affected in a major way and their lives are at risk. Both small- and large-scale textile industrial operations are considered to be significant contributors to the national economy in terms of production as well as employment and this sector was found to be the largest among the industries in the world. A large amount of waste is generated in the lengthy process followed in the textile industry due to the use of various resources, such as water, fuel, and different chemicals. The discharge of untreated effluents is a major issue associated with the textile industry, which creates water pollution issues and also other environmental issues that are equally as important, such as air emissions, especially volatile organic compounds (VOCs), odor, noise, and safety at the workplace (UNEP United Nations environment Programme, 1994; Vershney, 1985a).

2.1.1 Textiles in India

In India, the textile sector ranks next to the agricultural sector and is considered as one of the oldest industries in India. It has a persistent place in the national economy. Textile and clothing alone yields 27% of foreign exchange earnings. The textile and clothing industry contributes 14% of industrial production and 3% of gross domestic production. It provides 21% of the employment, with 35 million people employed directly in textile manufacturing activities and 60 million people indirectly employed, including cotton and jute growers, as well as artisans and weavers who are involved in the textile sector. The textile industry accounts for one-third of the total value of exports. The Indian textile industry depends greatly on cotton, with cotton making up 65% of the raw materials and jute products coming next. There are approximately 1227 textile mills with a spinning capacity of about 29 million spindles (Anand, 2014; Patel, 2016). Yarns are mostly produced in mills and fabrics are produced in power looms and in handlooms.

India stands second next to china for both cotton yarn and fabric production and fifth in the production of synthetic fibers and yarns. India's textile industry consists of small-scale nonintegrated spinning, weaving, finishing, and apparel-making enterprises (Inderjeet et al., 1991; Samanta et al., 2014).

1. *Spinning*: It is a process of converting cotton or manmade fiber into yarn, which is further used for weaving and knitting. In India's textile industry, spinning is considered to be the most efficient and consolidated sector.
2. *Weaving and knitting*: Conversion of cotton, manmade, or blended yarns into woven or knitted fabrics are done in weaving and knitting. India's weaving and knitting sectors are found to be highly fragmented, small scale, and labor intensive.
3. *Fabric finishing*: Dyeing, printing, and cloth preparation are included in fabric finishing, which is performed prior to the manufacture of clothing and it comes under small-scale enterprises. Around 2300 processors are operating in India, which includes 2100 independent units and 200 units with integrated spinning, weaving, or knitting units (Board, 2003).
4. *Clothing and apparel*: Around 77,000 small-scale units are classified as domestic manufacturers.

2.2 General process in dying industry

The textile industry begins with the harvesting or production of raw fiber and ends with the final products (Ministry of Textiles, 2012–2013; Shaikh, 2009; Pani, 2007). The various aspects of textile industries are schematically represented in Fig. 2.1.

2.2.1 Fabrics

The fabrics we use come in woven form, constructed by interlacing sets of yarn that run in both lengthwise and crosswise directions and with each yarn made up of several fibers. Fiber is defined as "any product capable of being woven or otherwise made into fabric." It is the smallest visible unit of a textile product. A fiber can be defined as a "pliable" hair-like strand that is very small in diameter in relation to its length. Fibers are considered to be fundamental units, or the building blocks, used in the making of yarns and fabrics. Both natural and manmade fibers are used in textile industries. Manmade fibers consist of both synthetic and regenerative cellulosic materials of petrochemical and wood fibers. A detailed description of fibers is given in Fig. 2.2.

Water and Textiles

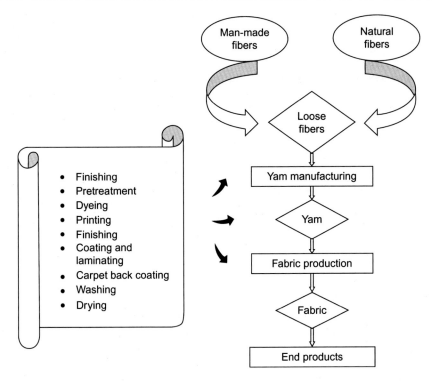

Fig. 2.1 Schematic representation of textile manufacturing industry.

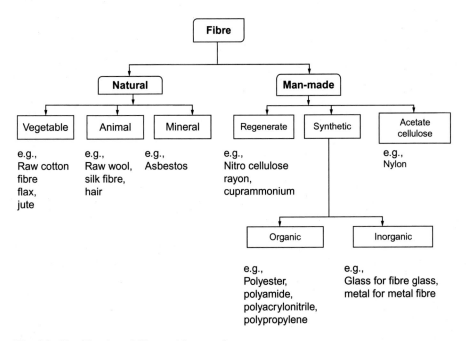

Fig. 2.2 Classification of fibers with examples.

2.2.2 Finishing processes

2.2.2.1 Pretreatment

The pretreatment process depends on the kind of fiber to be treated, the form of the fiber and the amount of material to be handled. Pretreatment is performed in order to remove foreign materials from fabrics and to improve the uniformity, hydrophilic characteristics, and affinity for dye stuffs. Generally, the pretreatment process is performed on the same equipment used for dyeing. Pretreatment is a must for removing unwanted materials from textiles. It is a series of cleaning operations in which all the impurities that cause adverse effects in dyeing and printing are removed. The pretreatment process differs according to the shade of the dyed material and depends on the buyer's requirement. After pretreatment, textile materials are ready for dyeing.

2.2.2.2 Dyeing

It is a process of coloring a textile material in which a dye is applied to the substrate in a uniform manner in order to obtain an even shade with greater performance. Four different steps are involved from a molecular point of view: (1) the dye diffuses from the liquor to the substrate in which the dye is previously dissolved or dispersed in the dye liquor; (2) accumulation occurs on the surface of the textile material; (3) the dye migrates into the interior surface of the fiber until uniformity is achieved; (4) the dye must be anchored to suitable places within the substrate.

In any stage of the manufacturing process, the textile can be colored and the following coloring processes are possible:

- *Flock or stock dyeing*: The staple of fibers can be dyed before spinning.
- *Top dyeing*: The fibers are shaped into lightly twisted roving before being dyed.
- *Tow dyeing*: This is the dyeing of monofilament material during the manufacture of synthetic fibers.
- *Yarn dyeing*: Yarn dyeing is different from other dyeing in that dyed yarns are used for making stripe knit or woven fabrics or in sweater manufacturing. Dyed in the form of package or hank form.
- *Piece dyeing*: The fabric is woven or knitted using greige (undyed) yarns, then the fabric is dyed.
- Ready-made goods.

Dyeing can be carried out either in continuous or batch mode. Dyeing depends upon the type of makeup, chosen class of dye, equipment availability, and the cost involved. The steps involved in both continuous and batch dyeing are as follows:

- Preparation of the dye;
- Dyeing;
- Fixation, washing, and drying;
- Printing. It is a process of application of color to a substrate. In order to obtain the desired pattern, color is applied to defined areas and the printing process involved is given below:

- Color-paste preparation. In this step, dyes or pigments are found to be in the form of finely dispersed in a printing paste under high concentration
- Printing. The dye paste is applied over the substrate using a variety of techniques.
- Fixation. After printing the fabric is dried and fixed with steam or hot air.
- After-treatment. Finally, the fabric is washed or dried.

2.2.2.3 Finishing

It involves mechanical/physical and chemical treatments that impart visual effects, improved handling, water proofing, and nonflammability. Finishing involves all the treatments from the textile to end-use products.

2.2.2.4 Washing

In the presence of a wetting agent and detergent, washing of textile material is usually carried out with hot water (40–100°C). The detergent used in the process emulsifies the oils and disperses undissolved pigments. Washing steps are included finally in order to remove emulsified impurities.

2.2.2.5 Softening

Softening as well as dry cleaning are sometimes done for delicate fabrics. Usually tetrachloroethylene is used for the dry-cleaning process. In the softening process, water and surfactant-based chemicals are added as solvent.

2.2.2.6 Drying

Elimination or reduction of the water content of fibers, yarns, and fabrics followed by wet processes. Drying by water evaporation is a highly energy-consuming step.

2.3 Water consumption in textiles

Our environment is a combination of all social, economic, biological, physical, and chemical factors. A symbiotic relationship exists between humankind and the environment. Our environment is found to be a complex and dynamic structure with interdependency of all life forms. The basic functions of mankind in relation to the environment are generally categorized into three and they are: (1) providing living space and other facilities that make human life qualitatively richer; (2) the environment acts as a platform in which abundant sources for agricultural, mineral, water and other resources are utilized directly as well as indirectly by humans; and (3) use of the environment as a sink in which all types of waste are disposed in one form or another. Humans are facing continuous potential health risks by disturbing the balance of nature and by changing the basic characteristics of the environment. Our environment includes everything around us, that is, air, water, and land. These are contaminated by the addition of chemical additives called "pollutants." Pollution is considered to be the introduction of contaminants to our environment. Water pollution may be defined as the addition of undesirable substances or unwanted foreign matter into water bodies, thereby adversely altering the natural quality of the water. Water is considered to be an important topic in environmental science and also considered to be an important substance in all parts of the environment as it covers 70% of the surface of the Earth. Water occurs in all spheres of the environment, in the oceans as a reservoir of saltwater, on land as surface water in rivers and in lakes, underground as groundwater, in the atmosphere as water vapor, and in the polar ice caps as solid ice. Water is considered to be essential for all living systems and assumed to be one of the mediums from which life has evolved and exists. Water demand is increasing considerably due to continuous economic development and growth in developed countries, as well as in developing countries. Because of the dwindling supply of good-quality water resources, it will be difficult to meet future demand. In a few decades, an imbalance in the demand for water and the supply of water resources will come up as a major problem. Physical, chemical, and biological interactions with the aquatic environment determine the water characteristics. Because of geochemical characteristics and geologic age, dynamic states of change have been seen in water bodies, such as rivers, lakes, and estuaries. This change is subjected to continuous accumulation and transformation of matter by living things and their associated activities and this results in pollution, which is seen in the odor and taste, as well as a reduction in the fish population. Water quality characterization includes an examination of: (1) chemical distribution dynamics in the aqueous phase (ice: soluble, colloidal, absorbed, or particulate matter); (2) chemical accumulation by aquatic biota; (3) accumulation of bottom deposits; and (4) land and atmosphere inputs. (Vershney, 1985b; De, 2006; Patel and Vashi, 2015; US EPA, 1996) Different types of machinery in the textile industry use different quantities of water. The water consumption of a batch processing machine depends upon the bath ratio (ratio of mass of water to the mass of fabric), physical flow characteristics, mechanical factors, such as agitation, mixing, and the bath and fabric turnover rate, and all these factors significantly affect the washing efficiency. Preferences are given to low-bath-ratio dyeing equipment as it conserves energy and water and achieves higher fixation efficiency (Lockerbie and Skelly, 2003). Three water quality types are suggested for use in textile industries (Water Treatment Solutions, 2010):

- *High-quality water.* In all phases of the textile industry, such as in dye baths, print pastes, and finishing baths, this type of water can be used. Compared with normal water, 10%–20% consumption of high-quality water is sufficient.
- *Moderate-quality water.* In washing-off stages, such as scouring, bleaching, dyeing, and finishing, this type of water is used. In the final stage of all the processes, high-quality water is used in order to remove finer contamination.
- *Low-quality water.* In general, washing of print paste containers, and the washing of screens and floors may use this type of water. Utilization of high-quality water is considered to be waste in this sector.

Dyeing effluents contain different varieties of pollutants, such as dispersants, leveling agents, salts, acids, alkali, and various dyes.

As discussed above, current wet processing operations followed in the textile industry are found to be chemical intensive and nonbiodegradable products are discharged in these operations. The toxic chemicals need special treatment to be removed from the wastewater. Most of the textile manufacturing industries are found in developing and less-developed countries and they lack equipped technologies, which results in the discharge of wastewater without any treatment, thus resulting in environmental contamination. The contamination has the ability to enter the food chain and such contamination is transferred to humans and to animals via bioaccumulation. The southern part of India is considered as a major textile processing area consisting of many small-scale textile plants. In a research project in and around Tirupur, discharges of untreated effluent in the river Noyyal over the past 20 years in which accumulation occurred in soil and in water, were observed. Groundwater, surface water, soil, and the aquatic ecosystem in Tirupur and the surrounding areas have been affected. Apart from wastewater, the burning of fossil fuels in the processing operation emits greenhouse gases, which implies global warming and climate change (Matioli et al., 2002; Uqaili and Harijan, 2011).

2.3.1 Impact of water resources

The effluents discharged from the textile processing industry make it into water bodies and water bodies are targeted first by the wet processing operations as water is used as a medium as well as solvent for the dissolution process, washing-off agent, and in generation of steam for heating the process bath. The quantity of water utilized depends upon the type of the fabric, the nature of the dye, the finishing agents, and the type of processing machinery used. For the processing of 1 kg of textile material, 50–100 L of water are required on average (Vershney, 1985b). When comparing natural fibers with synthetic fibers, natural fibers consume more water for the scouring and bleaching processes because water is required to remove the natural impurities in the fibers. A varying quality and quantity of complex mixtures of chemicals, both organic and inorganic, are present in these types of effluents. Chemical oxygen demand (COD), biological oxygen demand (BOD), total suspended solids (TSS), pH, total dissolved solids (TSS) values and low dissolved oxygen are commonly found in discharged effluent. Significant sources of pollution are seen. Desizing and scouring contributes 50% of the BOD in wastewater (World Bank, 2010; Mansour et al., 2012; Solmaz et al., 2011). Table 2.1 provides a comparison of drinking water standards as well as the water quality characteristics during the process of washing off and in washing down the equipment in the textile industry.

The process of dyeing generates a large quantity of wastewater. Dye preparation, the spent dye bath, and the washing process are the sources of wastewater in dyeing. Dyeing wastewater contains a large amount of salt, alkalinity, and a wide range of colors. Table 2.2 lists the dyeing processes with their water requirements.

Resin residues, softeners, and other auxiliaries generate organic pollutants in the finishing process. The water discharged from the process contains starches, dextrin, gums, glucose, waxes, pectin, alcohol, fatty acids, acetic acid, soap detergents, sodium hydroxide, sulfides, sulfites, chlorides, dyes, carbonates, pigments, carboxymethyl cellulose, gelatin, silicones, fluorocarbons, and resins (Gomez et al., 2008; Landage, 2009).

Table 2.1 Maximum water quality characteristics and water quality standards

Parameter	Indian standard for drinking water (BIS, New Delhi, India) (IS:10500, 1991)		Maximum water quality characteristics in textile industries	
	Requirements	Permissible limit	Washing off process	Equipment washing down
Color (Hazen units)	5		Not visible	Not visible
COD (mg L^{-1})	–	–	200	500–2000
Total hardness (ppm)	300	600	100	100
Chloride (mg L^{-1})	250	1000	500–2000	3000–4000
Fe (mg L^{-1})	0.3	1	0.1	0.1
Cu (mg L^{-1})	0.005	1.5	0.05	0.05
Cr (mg L^{-1})	0.05	No relaxation	0.1	0.1
pH	6.5–8.5	No relaxation	7–8	6.5–8.0

Table 2.2 Different types of dyeing processes with their water requirements

Different types of process	Water usage (mL kg^{-1})
ULLR dyeing	5000
Pad batch	5000
Package dyeing	5000–8000
Jet dyeing	7000–10,000
Hank machines	30,000

2.3.2 Characteristics of wastewater

The textile industry is considered to be among the most polluting industries as it produces a large amount of wastewater in different processes. The amount of wastewater in the form of cleaning water, process water, noncontact cooling water, and storm water depends upon various factors, including the nature of the processed fabric, the applied dye, and the type of process and equipment used. Processes such as scouring, dyeing, printing, finishing, and washing generate a large amount of wastewater. Wash water from continuous dyeing, alkaline wastewater, and batch dye wastewater contains dye, salts, acids, and other toxic additives. The amount of wastewater generated in the sizing and desizing operations followed by bleaching, dyeing, and printing is high when compared to other processes in the textile mill. In terms of raw water, around 61–646 L kg^{-1} of cloth are used in Indian mills. Fifty eight to eight one percent of wastewater is discharged from the entire amount of water consumed and an average value of 735 is discharged from wet processes. Wastewater is discharged at different steps in textile processing. The composition of wastewater depends on many factors, such as the type of fabric, the type of process, and the chemicals used. A large number of hazardous compounds are also emitted from the textile and dyeing industries. The characterization of wastewater for individual processes has been done and the wastewater from individual processes is found to be high in organic loads, color, and in all substances, which creates a negative impact to our environment (Elahee, 2010; Tufekci et al., 2007; Noyes, 1993; Bal, 1999; Yusuff, 2004; Humphries, 2009). Table 2.3 presents chemical uses and their effluent characteristics in different processes.

Table 2.3 Effluent composition and chemical usage in major processes in textile industries

Major processes	Effluent composition	Chemicals used	Nature
Desizing	Organic load (enzyme, acetic acid, starches) pH and salt effects (mineral/organic acids, carbonate) Refractory organics (synthetic sizes)	Starch, CMC, PVA, fats, waxes, pectin	High in BOD, COD, SS, DS
Scouring	pH and salt effects (Na OH, Na_2Co_3) Organic load (oil, fats, wax)	Starch, waxes, carboxymethyl cellulose, polyvinyl alcohol, wetting agents	High in BOD, COD
Bleaching	Organic load (NaOCl, $NaClO_2$) pH and salt effects (sodium chloride) Toxicants (heavy metals, reducing agents and oxidizing agents) Refractory organics (carrier organic solvents, chlorinated organic compounds)	Sodium hypochlorite, NaOH, acids, sodium phosphate, cotton fiber	High in alkalinity and in SS
Mercerizing	pH and salt effects (NaOH, NaCl)	Sodium hypochlorite	High in pH, high in DS, high BOD

Continued

Table 2.3 Continued

Major processes	Effluent composition	Chemicals used	Nature
Dyeing	Organic load (enzymes, surfactants, acetic acid) Color (dyes, scoured wool impurities) Nutrients (ammonium salts, urea phosphate–based buffers and sequestrants) Sulfur (sulfate, sulfide, hydrosulfide, salts, and sulfuric acid) Refractory organics (dyes, resins) PH and salt effects (silicate, sulfate) Toxicants (metals)	Urea, reducing agents, acetic acid, wetting agents	High in BOD, DS, SS, low heavy metals
Finishing	Refractory organics (surfactants, synthetic sizes) Toxicants (reducing agents e.g., sulfide, chlorinated compounds), oxidizing agents (e.g., chlorite, peroxide, dichromate)	Pastes, gums, oils, thickeners, reducing agents	High in BOD, suspended solid, slightly alkaline

2.4 Water consumption in life cycle phases of textile products

Textile industries are similar to chemical industries and they cause pollution in a similar or in a different pattern from chemical industries. A large amount of water is consumed by textile industries at various stages. Except for spinning and weaving, every other process utilizes water extensively. All dyes and chemicals are applied to fabric substrates in water baths. An aqueous system is used in fabric preparation steps, including desizing, scouring, bleaching, and mercerizing. According to USEPA, 36,000 L of water are utilized for the production of 20,000 Lb day^{-1} (VolmajerValh et al., 2011). A million gallons of dye wastewater are generated in textile mills and unnecessary usage of water adds cost to the finished textile products. The quantity and usage of water depends upon the fabric produced, the process equipment type, and the dyestuff. For longer processing sequences, the quantity of water required is greater.

There are five phases in textile production. Various processes and materials are involved in each phase from agriculture and farming to chemical usage and mechanical operations. The textile production process from fiber to garment production is described below. Fig. 2.3 shows the sources of water pollutants from various processes and Table 2.4 presents the amount of water utilized by different processes and various

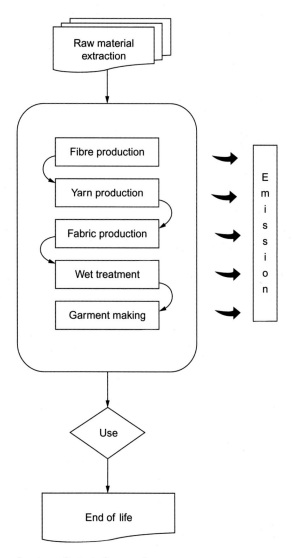

Fig. 2.3 Sources of water pollutants from various processes.

Table 2.4 Water use with various fibers and processes in the textile industry

Process in textile industry	Water use for different fibers in each process (mg kg^{-1})						
	Cotton	Wool	Rayon	Acetate	Nylon	Acrylic	Polyester
Sizing	500–8200						
Desizing	2500–21,000						
Scouring	20,000–45,000	46,000–100,000	17,000–34,000	25,000–84,000	50,000–67,000	50,000–67,000	25,000–42,000
Salt bath	–	–	4000–12,000	–	–	–	–
Washing	–	334,000–835,000	–	–	–	–	–
Neutralization	–	104,000–131,000	–	–	–	–	–
Bleaching	2500–25,000	3000–22,000	–	33,000–50,000	–	–	–
Mercerizing	17,000–32,000						
Dyeing	10,000–300,000	16,000–22,000	17,000–34,000	34,000–50,000	17,000–340,000	17,000–34,000	17,000–34,000
Printing/ special finishing	8000–16,000		4000–12,000	24,000–40,000	32,000–48,000	40,000–56,000	8000–12,000

fabrics. This section provides detailed information about the phases involved in textile products as well as their water consumption (Roos, 2016; Isik and Sponza, 2008).

2.5 Fiber production phase

Fiber production and use are increasing gradually due to population growth. Fibers are classified mainly into two categories and they are natural fibers, such as cotton, wool, silk, and flax, and synthetic fibers, which are mainly obtained from fossils and include polyester, polyamide, and elastane. Cotton is found in fiber form and it is harvested manually or by machine with separation of lint and seed. About 20 million tons of cotton are produced in around 90 countries and half of the textile cloths made worldwide use cotton with the rest coming from synthetic fibers. One kilogram of cotton consumes approximately 20,000 L during its production (Barlocher et al., 1999). Wool can be obtained from sheep, goats, alpacas, camels, rabbits, and other animals. Once the impurities are removed, the wool is ready to spun and dyed. Water is required during the cultivation and farming stages. Synthetic fibers are manufactured from plastic or pulp and they are converted to fiber shape via fiber spinning. Three types of fiber spinning technologies are currently in process and they are wet spinning, melt spinning, and dry spinning.

2.5.1 Yarn production phase

In this process, the fibers are drawn into filament yarns or into staple fibers depending upon the need. Generally, synthetic fibers are drawn to filament yarn in order to produce spun yarn and natural fibers are directly harvested as spun yarn and staple fibers. Sinning is achieved by a twisting step so that the produced yarn is found to be pressure resistant during weaving. Fiber thickness and filament numbers are included while fixing the yarn size. In subsequent steps, the consumption of energy and chemicals is determined by the size of the yarn.

2.5.2 Fabric production phase

Fabrics come in woven and nonwoven forms. Fabrics are produced from gray yarns (yarn which is not bleached). In order to attain good strength, yarns are sized using lubricators. The weave produced must be desized using washing and drying in a wet treatment process. There is a knitting process in fabric production. Two types of knitting are followed in this process and they are: (1) circular knitting using high gauges, and (2) flat knitting for low gauges. Nonwovens are directly produced from staple fibers or from filaments. Fabric strength is increased by bonding with resin or needle-punching.

2.5.2.1 Wet treatment

The whole process of dyeing can be classified into two processes and they are the dry process and the wet process.

Dry process: The dry process consists of opening, blending, mixing, carding, combing, spinning, weaving, and knitting and the water use in this process is found to be negligible.

Wet process: Water is used for solvent processing and for the washing and rinsing medium. The wet process consists of singeing, desizing, kiering, bleaching, mercerizing, and dyeing. In each stage, an appreciable amount of water is necessary for the series of operation (Rajagopalan, 1990; Pandey and Carney, 2008; Bisschops, 2003; Patel and Vashi, 2015).

- *Singeing*: In this step, the fabric travels near a series of jet burners at a fast rate for removing protruding fibers from the surface. In order to make the fabric surface smooth, flames are made to burn over the surface, which results in the burning of fibers and thus a smoother surface. The addition of starch or other nonbiodegradable sizing agents strengthens the fibers. The wastewater consists of starch and softeners, which originate from the washing processes of various vessels. Instead of starch, carboxymethyl cellulose (CMC) and polyvinyl alcohol (PVA) are also used.
- *Desizing*: Desizing is the next step after singeing. It involves removing natural impurities and singeing compounds. Generally, water soluble sizes are used for man-made fibers with a hot-water wash or scouring process. In this step, different types of enzymes are used for hydrolyzing starch. Removal of starch before the kiering process is necessary as it reacts with the sodium hydroxide in the kier process. In desizing the fabric, wastewater occurs with a high concentration of organic material, consisting of breakdown products, such as sizing materials and hydrolysis agents.
- *Kiering*: Kiering is done to both natural as well as synthetic materials. After the process of desizing, the fibers contain grease, lubricants, antistatic agents, waxes, etc. By scouring with alkaline liquor containing caustic soda, soda ash, sodium silicates, sodium peroxide, or others for several hours under steam, the impurities are removed further. This occurs in batch and the spent liquor is blown out instantly. Water scouring is preferred over solvent scouring. For cotton scouring, an alkaline solution containing soaps or detergents is used. The scouring effluent can contain herbicides, insecticides, defoliants, and desiccants, which are used in the production of cotton, and also fungicides, such as pentachlorophenols. Apart from cotton, raw wool scouring is considered to be the most polluting in the industry. This is due to the wax, urine, feces, vegetable matter, mineral dirt, soap detergent, and alkali that appear in the scouring and washing process.
- *Bleaching*: The fabric is treated with bleach liquor. The liquor used is mostly hydrogen peroxide with other chemicals, such as sodium chloride, formic acid, sulfuric acid, caustic soda, and hypochlorites, used for removing natural coloring material. In bleaching, fabrics are first rinsed with water and then with dilute acid and sodium bisulfate to remove the trace amounts of chlorine and alkali. In order to improve the whiteness, the fabric is soaped, washed, and treated with optical whitening agent. In the whole amount of wastewater, this process contributes 10–205 L of wastewater. Hydrogen peroxide present in the wastewater reduces the pollution load, which results in less solid residual. The wastewater released from this process consists of soap and optical whitening agent.
- *Mercerizing*: Countercurrent system with water and with a treatment of cold concentrated caustic alkali solution. The fabric tensile strength, luster, and dye affinity are all increased in this step.

- *Finishing*: The unfixed dyes are removed in this step by washing the dye in an open soaping range and treating with starches, dextrins, natural and synthetic waxes, and synthetic resins in order to finish the fabric preparation process. The main water consumption occurs in washing the fabric for removing unused color and cleaning the color machine. The water from this step is strongly colored and contains some fixing agents, such as gum, soap, minerals, neutralizers, softeners, water-repellents, and flame retardants.
- *Garment making*: Printing is used in the garment-making sector, particularly in the sewing factory or in the dye house. Roller printing, screen printing, transfer printing, and others are all used. The dye stuffs for printing depends on the fiber material. In recent years, digital printing has increased, leading the way to pattern printing. In the garment-making sector, cutting, sewing, ironing, and packaging are also done.

2.6 Challenges and opportunities

Developments in fashion and restrictions in the criteria for wastewater discharge are the major reasons for the reuse of process water and chemicals. Determining the general standard quality for textile wastewater remains a challenge as different types of fiber have different requirements on the water in the final stage (NC Department of Environment, Health and Natural Resources, 1993). Prevention of pollution is considered to be great challenge and is not available in most sections of the industry. Fig. 2.4 sketches the textile industry process impact on the environment.

Companies, whether small scale or large scale, should employ a start-of-pipe approach as well as an end-of-pipe approach. In order to increase the process efficiency, increase company profits, and to minimize the impact to our environment, pollution-prevention techniques should be implemented. The understanding of pollutant release and its synergistic effect on air, land, and water should be studied carefully (Pani, 2007; Uqaili and Harijan, 2011).

Some of the known pollution prevention techniques to be adopted in the textile industry at the manufacturing level are as follows,

- *Quality control in raw materials.* The introduction of less-polluting raw materials is necessary and, in order to avoid waste production, prescreening of raw materials has to be done.
- *Less hazardous chemical substitution.* The textile industry is considered to be use chemical intensive processes and substitution of chemicals depends on environmental and process conditions, products, and raw materials. The substitution of chemicals is possible in the areas of desizing and of dyes and their auxiliaries.
- *Modification in process.* Modification of operations is necessary in order to attain optimization in the process, for example, the substitution of dyeing machines which have a low liquor ratio and combining operations in order to save energy and water.
- *Equipment modification.* Complete automation of the process helps in reducing the amount of waste and human error resulting from the tediousness of the conventional process.

2.6.1 Water conservation and reuse

The textile industry is considered to be very water intensive with water being used to clean the raw materials as well as being used in many flushing steps during production.

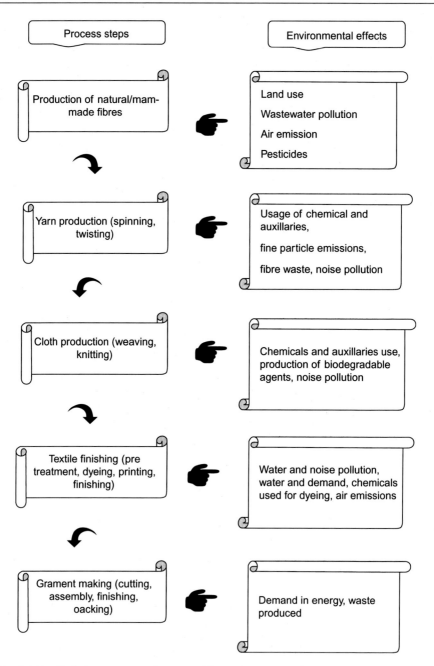

Fig. 2.4 Textile industry process impact and its environmental effects.

Water conservation and reuse are becoming a necessity for the textile industry with benefits, such as a decrease in the cost of purchased water and a reduction in the cost for the treatment of wastewater. The wastewater has to be cleaned of fat, oil, color, and other chemicals which are utilized in the various processing steps.

The first step in water conservation and reuse is to conduct a site survey and to develop a detailed spreadsheet of water usage with specific details. Such details include:

- Quantity and location of water usage,
- Water quality requirement, that is, pH, hardness, temperature, etc.,
- Requirement for specific processes, etc.

Listed below are some of the techniques used for water conservation:

- *Counter-current washing.* For dyeing, counter-current washing is employed frequently. In this technique, clean water flows counter to the movement of the fabric through the wash boxes. It is designed in such a way that the contaminated wash water contacts the fabric first and later the fabric contacts the cleanest water.
- *Reuse of final rinse water from dyeing for dye bath makeup.* The rinse water from the final rinse of the batch dyeing operation is treated and can be used for further rinsing or to make up dye baths.
- *Reuse of soap wastewater.* The wastewater obtained from the soaping operation can be used for cleaning the floors and equipment and can also be reused at the backgrey washer.
- Reuse of bleach wash water for scouring or desizing.
- *Use of automatic shutoff valves.* Thermally controlled shut-off valves for each process unit have to be provided.
- *Use of flow control valves.* This type of valve is used in cleaning areas in which water conservation is important.
- *Good housekeeping.* By taking strict housekeeping measures, from 10% to 30% of water can be conserved. This includes auditing broken or missing valves, leakage from pipes, valves, and pumps, defective toilets, and water coolers, and excessive water utilization in washing operations.
- Good housekeeping leads to savings in cost, water, chemicals, and energy. Good housekeeping is unavoidable for the growth of a company (Water Treatment Solutions, 2010; Bergenthal, 2004; Shaikh, 2009).

Two types of approaches are highlighted for water reuse and recycling and they are the start-of-pipe approach and the end-of-pipe approach, with the approach determined by industrial need.

2.6.1.1 Start-of-pipe approach

In order to save water, chemicals, and energy, the dye bath should be reused for further dye treatments by knowing the present characteristics of the dye bath. The water networks and water treatment techniques have to be analyzed and optimized. Several tools have been identified to determine optimal designs for energy and mass-transfer networks. Tools, such as pinch analysis, provide a way to identify ideas for reuse, regeneration, and water treatment. This tool has been found to be reliable, efficient, and cost-effective for the reuse of water (US EPA, 1996; Matioli et al., 2002).

2.6.1.2 End-of-pipe approach

This method is well suited to the case in which many textile factories (medium or small enterprises) use the same water facilities. The end-of-pipe approach especially deals with wastewater produced in multiple stages and it uses biological and physicochemical techniques (Schoeberl et al., 2004).

2.7 Conclusion

The textile industry is considered to be among the oldest and largest industries worldwide. Among all the processes in the textile industry, wet processing is responsible for the utilization of a large amount of resources, in terms of both water and energy. Through consumption of fresh water and the discharge of untreated water to the environment, the textile industry has a significant impact on the aquatic ecosystem. Due to lower availability of fresh water, water reclamation and reuse are considered to be of economic interest. The use of high-quality water in all production processes is a preferable option for small-scale industries as it results in less water use throughout the process. In the last decades, innovation in terms of new technologies for the treatment of wastewater has occurred, but the technologies are often restricted in efficiency and cost. Textile-related activities in India are mostly comprised of small-scale enterprises in which separate treatment of wastewater or used water is not possible. In recent times, due to climactic change, the world has been facing water-related threats as climate change is influencing the hydrological cycle. Conservation of fresh water is thus mandatory for today's scenario. Optimization in process operation, design, and equipment is necessary for the efficient use of fresh water. Research and development should be performed in the area of biodegradable dyes, chemicals, and their auxiliaries. The conservation of water resources lies in the hands of concerned industries and researchers through the development of alternative techniques that cause minimum impact to our environment. Sustainability in textile production has to be developed and government should play a key role by tightening regulations for the implementation of sustainable textile production for the purposes of ensuring a clean environment and livelihoods for people.

References

Anand, M., 2014. A study of financial analysis in textile sector. Int. J. Bus. Manag. Soc. Res. 3 (6), 2319–5614.
Bal, A.S., 1999. Wastewater management for textile industry—an overview. Indian J. Environ. Health 41 (4), 264–290.
Barlocher, C., Holland, R., Gujja, B., 1999. The Impact of Cotton on Fresh Water Resources and Ecosystems. WWF International, Gland.
Bergenthal, J.F., 2004. Wastewater recycle and reuse potential for indirect discharge textile finishing mills, volume 1. Environmental Protection Agency, Technical report. U.S.

Bisschops, H., 2003. Spanjers, Literature review on textile wastewater characterization. Environ. Technol. 24 (11), 1399–1411.
Board, N., 2003. The Complete Technology Book on Textile Spinning, Weaving, Finishing and Printing. Asia Pacific Business Press, New Delhi.
De, A.K., 2006. Environmental Chemistry, sixth ed. New Age International Publishers, New Delhi.
Elahee, K., 2010. Heat recovery in the textile dyeing and finishing industry: lessons from developing economies. J. Energy South Afr. 21 (3), 9–15.
Gomez, N., Sierra, M.V., Cortelezzi, A., Rodrigues Capitulo, A., 2008. Effects of discharges from the textile industry on the biotic integrity of benthic assemblages. Ecotoxicol. Environ. Saf. 69 (3), 472–478.
Humphries, M., 2009. Fabric Reference, fourth ed. Pearson Education, Inc., Upper Saddle River.
Inderjeet, S., Sethi, M.S., Iqbal, S.A., 1991. Environmental Pollution: Cause, Effects and Control. Common Wealth Publishers, New Delhi.
Isik, M., Sponza, D.T., 2008. Anaerobic/aerobic treatment of a simulated textile wastewater. Sep. Purif. Technol. 60 (1), 64–72.
Landage, S.M., 2009. Removal of heavy metals from textile effluent. Colourage 56, 51–56.
Lockerbie, M., Skelly, J.K., 2003. Water Quality Requirements for Treated Effluent Produced by Recycling. Water Recycling in Textile Wet Processing, SDC, Bradford, pp. 212–223.
Mansour, H.B., Houas, I., Montassar, F., Ghedira, K., Barillier, D., Mosrati, R., Chekir-Ghedira, L., 2012. Alteration of in vitro and acute in vivo toxicity of textile dyeing wastewater after chemical and biological remediation. Environ. Sci. Pollut. Res. 19 (7), 2634–2643.
Matioli, D., Malpei, F., Bortone, G., Rozzi, A., 2002. Water minimisation and reuse in the textile industry. In: Lens, P., Pol, L.H., Wilderer, P., Asano, T. (Eds.), Water Recycling and Resource Recovery in Industry, Analysis, Technologies and Implementation. IWA Publishing, London, pp. 545–584.
Ministry of Textiles, 2012–2013, Government of India (various years) *Annual Report*, New Delhi: Ministry of Textiles.
NC Department of Environment, Health and Natural Resources, 1993. Waste Reduction Fact Sheet, up-to-the-Minute Waste Reduction Techniques and Technologies, NC Division of Pollution Prevention and Environmental Assistance. NC Department of Environment, Health and Natural Resources. From: http://infohouse.p2ric.org/ref/01/00026.htm.
Noyes, R., 1993. Pollution Prevention Technology Handbook. Noyes Publications, New Jersey.
Pandey, G.N., Carney, G.C., 2008. Environmental Engineering. Tata McGraw-Hill Publishing, New Delhi.
Pani, B., 2007. Textbook of Environmental Chemistry. I.K. International Publishing House Private Limited, New Delhi.
Patel, M.V., 2016. Comparative study of ratio analysis of selected textile companies of India. Int. J. Humanit. Soc. Sci. 4 (3), 43–48.
Patel, H., Vashi, R.T., 2015. Characterization and Treatment of Textile Wastewater. Elsevier.
Rajagopalan, S., 1990. Water pollution problem in textile industry and control. In: Trivedy, R.K. (Ed.), Pollution Management in Industries. Environmental Pollution, Karad.
Roos, S., 2016. Advancing Life Cycle Assessment of Textile Products to Include Textile Chemicals. Inventory Data and Toxicity Impact Assessment. Thesis for the degree of doctor of philosophy, Gothenburg, Sweden.

Samanta, K.K., Basak, S., Chattopadhyay, S.K., 2014. Muthu, S.S. (Ed.), Roadmap to Sustainable Textiles and Clothing, Textile Science and Clothing Technology. Springer, Singapore, pp. 263–287.

Schoeberl, P., Brik, M., Braun, R., Fuchs, W., 2004. Treatment and recycling of textile wastewater—case study and development of a recycling concept. Desalination 171, 173–183.

Shaikh, M.A., 2009. Water conservation in textile industry. Pak. Text. J. 48–51. From: https://www.ptj.com.pk/Web-2009/11-09/Muhammad-Ayaz-Shaikh.htm.

Solmaz, S.K., Birgul, A., Ustün, G.E., Yonar, T., 2011. Colour and COD removal from textile effluent by coagulation and advanced oxidation processes. Color. Technol. 122 (2), 6102–6109.

Tufekci, N., Sivri, N., Toroz, I., 2007. Pollutants of textile industry wastewater. Turk. J. Fish. Aquat. Sci. 7, 97.

UNEP (United Nations environment Programme), 1994. The Textile Industry and the Environment. Technical Report No. 16. UNEP/IE, Paris.

Uqaili, M.A., Harijan, K. (Eds.), 2011. Energy, Environment and Sustainable Development. Springer Science & Business Media, New York.

US EPA, 1996. US EPA/625/R-96/004, Best Management Practices for Pollution Prevention in the Textile Industry, Manual. US Environmental Protection Agency, Cincinnati, OH.

Vershney, C.K. (Ed.), 1985a. Water Pollution and Management. New Delhi, Wiley Eastern.

Vershney, C.K. (Ed.), 1985b. Water Pollution and Management. Wiley Eastern, New Delhi.

VolmajerValh, J., Majcen Le Marechal, A., Vajnhandl, S., Jeric, T., Simon, E., 2011. Water in Textile Industry. University of Maribor, Maribor, Slovenia. Elsevier. Chapter 4.

Water Treatment Solutions, 2010. Water Recycling in the Textile Industry. http://www.lenntech.com/water_reuse_textile_industry.htm. Accessed April 2010.

World Bank, 2010. A Detailed Analysis on Industrial Pollution in Bangladesh. Workshop Discussion Paper Transfer, EPA-625/3-74-004. World Bank Dhaka Office, Dhaka.

Yusuff, R.O., Sonibare, J.A., 2004. Characterization of textile industries effluents in Kaduna, Nigeria and pollution implications. Global Nest Int. J. 6 (3), 212–221.

Water consumption in textile processing and sustainable approaches for its conservation

Kartick K. Samanta*, Pintu Pandit[†], Pratick Samanta[‡], Santanu Basak[§]
*Mechanical Processing Division, National Institute of Research on Jute and Allied Fibre Technology, Kolkata, India, [†]Department of Fibres & Textile Processing Technology, Institute of Chemical Technology, Mumbai, India, [‡]Department of Textile Technology, Indian Institute of Technology-Delhi (IIT-Delhi), New Delhi, India, [§]Chemical & Biochemical Processing Division, Central Institute for Research on Cotton Technology, Mumbai, India

3.1 Introduction

The textile industry is one of the major consumers of water and it ranks among the top ten water-consuming industries. Water is used in large quantities in different textile processing operations, starting from preparation, and continuing with sizing, desizing, scouring, bleaching, dyeing, printing, and value-added aesthetic and functional finishing. The amount of water used in the textile industry varies widely, depending on the specific processes operated at the factory, the equipment used, and management practice concerning the use of water. The water consumption in a process house is about three times the consumption of all other units put together. As far as the different natural textiles are concerned, the processes for wool and similar felted fabrics require more water consumption than other textile processing, such as that for woven, knits, stock, and carpet. As far as cotton textile is concerned, 95% of the water is used for raw material production and the remaining 5% is used in the processing of the cotton fabric. An estimated 100–150 L of water is needed to process 1 kg of cotton textile. Conventionally, to produce a pair of cotton jeans requires a larger amount of water (more water footprint). Protein fibers, such as wool and silk, also require greater amounts of water, mainly owing to scouring and coloration. Recently, it has been stated that a cellulosic fabric bleaching requires more water than other processes. As far as the detailed dyeing process is concerned, different amounts of water are needed for different classes of dyes. Furthermore, the water use also varies according to the dyeing machine used and the processes that are followed. Instead of textile coloration by dyeing, coloration by printing technology has also been considered, in which the water consumption is comparatively much less. Along these lines, the invention of digital printing is worth mentioning, it being an almost water-free process. In industry, it has been observed that the water use is almost three times greater in beam and beck dyeing machines compared to jigger dyeing machines; this may be attributed to a process with a high material-to-liquor ratio. Recently, different textile industries have been using continuous dyeing and the pretreatment range (pad-dry-pad-steam) for further reductions in the quantity of water needed and also for mass production. The application of an ultrasound-assisted process and supercritical fluids have also been attempted in the textile processing arena so as to reduce the water footprint. In the last few decades, the world has been moving toward the development of water saving processes due to a scarcity of clean water. Effluent generated from textile factories is another important concern that needs to be addressed properly. This has also been considered in the past either to establish a suitable treatment plant to partially or fully utilize the processed water or discharge of contamination-free water downstream, without affecting agriculture and aquaculture. The effluent generation problem increases the biological oxygen demand (BOD), chemical oxygen demand (COD) and total dissolved solid (TDS) level of the water used due to the presence of the different kinds of environmentally unfriendly dyes, pigments, finishing chemicals, alkali, acid, dispersing agents, and many more besides. In this regard, research and development has been attempted to validate the possibility of plasma induction, laser and UV irradiation, gamma irradiation, and ozone treatment in the processing of natural and synthetic textiles for the improvement of dyeing, printing, adhesion, cleaning, denim fading, felting, effluent treatment, hydrophilic and hydrophobic finishing, and many more (Samanta et al., 2014a, 2015a). Many such valued additions in textiles have also been attempted through various aqueous and solvent-based coatings (normal to foam) with varying thickness from the micron to submicron range; this has also led to a reduction of the water footprint. Furthermore, it helps in the production of functional textiles with a desired functionality by modifying either the bulk or surface of the material at the micron to nanometer scale. Indeed, these kinds of irradiation techniques could improve the rate of dyeing and the saturation dye uptake in natural as well as synthetic fibers in a much shorter interval.

The present chapter discusses the water required in the processing of different textiles, such as woven, nonwoven, knitted, and yarn. It also describes the water footprint in the different wet chemical processing steps of textiles, such as sizing, desizing, scouring, bleaching, dyeing, finishing, and washing. The development of textile dyeing and wet-chemical processing machinery has also been summarized from the point of view of the material: liquor (water) ratio and the total water requirement. The water requirements for processing cellulosic materials, cotton, lignocellulosic materials, jute, proteinous materials, wool, silk, and the manmade fibers nylon, acetate, and polyester have also been highlighted in details. Different strategies for the conservation of water and chemicals and also cost reduction have been discussed from the point of view of process development, machine development, and advances in application technologies. Toward the end, the application of various waterless or less-water processing technologies in textiles, such as plasma, UV, laser, ultrasound, and supercritical fluid, are presented from the point of view of water, energy, and chemical requirements, as well as cost and environment pollution.

3.2 Fiber, dye and process wise water requirement

The actual quantity of water required for the textile wet processing depends upon the type of material being processed, the nature of the dyes, the finishing agent, and the type of machinery used for processing. On average, 50–100 L of water (world average) is required for processing 1 kg of textile material (Uqaili and Harijan, 2011). In textile wet processing, water is essential in almost every stage along with suitable chemicals.

The textile industry consumes a large amount of water in various processes and an almost equal amount of water is discharged at the end of the processing cycle as the amount of water used as a processing medium. Consumption of water in the mechanical processing (spinning and weaving) is minimal compared with textile chemical processing (scouring, bleaching, dyeing, printing, finishing), for which water is essential. Almost all of the dyes and chemicals applied to textile substrates are applied in aqueous media. According to The United States Environmental Protection Agency (USEPA) a unit producing 20,000 lb/day of fabric consumes 36,000 L of water. In textile wet processing, water is used mainly for chemical processing of textile and as a medium for textile washing and rinsing. In the textile industry, some amount of water is also needed for running the boiler, cooling water, steam drying and for cleaning purposes (Shaikh and Ayaz, 2009; Rathore, 2012; Alexandra, 2009). Commonly, effluents discharged by the textile industry contain Polybrominated Diphenyl Ethers (PBDEs), phthalates, organochlorines, lead, and a host of other chemicals that have been proven to cause a variety of health issues (Tenenbaum and David, 1998). Water consumption in textile wet processing varies greatly, as shown in Table 3.1. This table summarizes the water requirements in various processing categories. Processing of felted fabrics is more water intensive compared with the processing of other types of fabrics, such as woven, knitwear, carpet, nonwoven, and stock.

Table 3.1 Water consumption in wet processing of different textiles[1]

Type of fabric	Water use minimum, gal/lb of production	Water use median, gal/lb of production	Water use maximum, gal/lb of production
Woven	0.6	13.6	60.9
Knit	2.4	10.0	45.2
Carpet	1.0	5.6	19.5
Stock/yarn	0.4	12.0	66.9
Nonwoven	0.3	4.8	9.9
Felted fabrics	4.0	25.5	111.8

3.2.1 Water consumption in processing of natural fibers

A large amount of water is consumed by the textile industry for processing natural fibers, such as cellulose (cotton), lignocellulose (jute, linen, hemp, banana), and proteinous materials (wool, silk). In the cotton textile industry, the sizing of warp yarn requires a good amount of water prior to other chemical processing, such as desizing, scouring, bleaching, dyeing, printing, and finishing. Furthermore, water is also essential in rinsing and washing of the processed fabrics. In the aforesaid processes, the water footprint varies widely, depending on the specific processes operated at the mill, the processing equipment used, and the prevailing management philosophy concerning water use. The approximate water consumption in the processing of the natural fiber cotton is 250–350 (kg/kg of fabric) and for wool is 200–300 (kg/kg of fabric)[1]. Cotton fibers consume a huge amount of water in their preparation. In general, the water consumption in a process house is about three times the consumption of all other input. Table 3.2 summarizes the water requirements in wet processing of textiles starting from desizing to bleaching, as well as the water requirements of the associated washing steps. It is seen in the case of knit processing mills that an average of 10 gallons of water per pound is required for production, though water use ranges from as low as 2.4 gallons to as high as 45.2 gallons.

It has been observed that, in practice, the quantity of water required for textile processing is large. However, it varies from mill to mill depending upon various factors, such as the type of fabric, equipment type, process, and dyestuff used. At the end of each process, the washing unit consumes a large quantity of water, as shown in Table 3.2. The processing of natural yarns (cotton, jute, sisal, etc.) also requires a large volume of water. The use of old and new technologies leads to a wide variation in water consumption in the processing steps as does the adoption of newer types of machinery, such as continuous ranges of dyeing and the pretreatment process. When processing natural fibers, the material-to-liquor (water) ratio varies with the type of dyeing machine used in wet processing. A low-liquor-ratio dyeing machine could save water, chemicals, and also energy to a great extent. The typical water requirements in dyeing along with the liquor ratios for various types of dyeing machines are shown in Table 3.3.

Table 3.2 Consumption of water and energy in J-Box[1]

Process	Consumption of water (L/kg)	Consumption of steam (kg/kg of fabric)
Desizing	3	0.25
Washing	20	0.35
Scouring	2	1.75
Washing	20	0.30
Bleaching	2	1.00
Washing	40	0.60
Total	87	4.20

[1] http:/textilelearner.blogspot.com/2014/04/water-consumption-in-textile-industry.html

Table 3.3 Material-to-liquor ratios and the water requirements in various equipment (Shaikh and Ayaz, 2009; GG 62, 1997; Saxena et al., 2017; Shukla, 2007; Ahmed et al., 2011)

Dyeing machine	Typical liquor ratio [liquor/goods during dyeing]
Continuous	1:1 [Vat dye—4.1 L/kg]
Winch	15:1–40:1
Jet	7:1–15:1
Jigger	5:1 [Direct dye—7.1 L/kg; Reactive dye—16.1 L/kg]
Beam	10:1
Package	10:1
Beck	17:1 [Direct dye—18.1 L/kg; Reactive—37.1 L/kg]
Stock	12:1
Skein	17:1

It has been established from the data that, compared with batch processing, continuous bleaching and dyeing machines are more efficient in terms of water consumption. Modern continuous textile processing machines, used by various textile industries are very economical in terms of water requirements. Sometimes, installation of such water-efficient machines is a concern due to the high capital investment and small running lengths in one processing shade, and also because they are unsuitable for fabrics that need tensionless handling, such as knitwear and crepe fabrics (Saxena et al., 2017).

3.2.2 Water consumption in processing of synthetic fibers

Table 3.4 presents the tentative water consumption and total water requirement in the processing of various manmade fiber-based textiles. In the synthetic textile industry, water consumption is lower than in the natural textile processing industry. Only 100–200 kg/kg water is required for wet processing of polyester. A wide variation in water consumption has been observed, mainly because of the use of old and new technologies, the type of fiber, and the type of machinery. The volume of water required in each process is also shown in Table 3.4.

3.3 Strategies for reduction of water consumption in textile processing

In textile processing, a large amount of water is used to impart color or finish in the textile substrate. Additionally, a large amount of energy is also required to dry the wet textile and run the processing machines. Many times, contaminated water is discharged without any effluent treatment, thus, the level of water pollution is enhanced. The generation of effluent is one of the biggest concerns in the textile industry and needs to be addressed properly. Some of the technological advancements in this direction, that is, to minimize effluent generation in textile wet processing, are dyeing with

Table 3.4 Water requirements for synthetic textile wet processing (World Bank, 2010; Mansour et al., 2012; Solmaz et al., 2006; Gómez et al., 2008)

Process	Water requirements in L/1000 kg of product made of different fibers				
	Rayon water	Acetate water	Nylon water	Acrylic/Modacrylic water	Polyester water
Scouring	17000–34000	25000–84000	50000–67000	50000–67000	25000–42000
Salt bath	4000–12000	–	–	–	–
Bleaching	–	33000–50000	–	–	–
Dyeing	17000–34000	34000–50000	17000–34000	17000–34000	17000–34000
Special Finishing	4000–12000	24000–40000	32000–48000	40000–56000	8000–12000

ultrasonic waves, microwaves, and with the electrochemical method, pretreatments using plasma technology, waterless dyeing with supercritical fluid, biotechnological catalytic bleaching, a single-stage preparatory process, foam finishing, and many more (Lloyd and Helmer, 1991; Teli, 1996; Lakshmanan and Raghavendran, 2017; Raichurkar and Ramachandran, 2015; Seshama et al., 2017; Teli et al., 2017; Teli and Pandit, 2017; Basak et al., 2016a).

3.3.1 Developments in dyeing machinery and processes to reduce water consumption

Textile manufacturers are trying to develop processing machines that not only improve the quality of the product, but also reduce the processing cost by minimizing energy use, reducing effluent generation and employing high-speed production. In the last few decades, machinery manufactures have been more focused on machines, such as CDR (continuous drying range) and CBR (continuous bleaching range) systems. Such processes provide high product quality with high efficiency. An airflow machine has been developed to reduce the material-to-liquor ratio to 1:3. For the finishing of synthetic materials, a closed high-temperature and high-pressure (HTHP) jigger machine has been developed to process the polyester fabric in opened width form with very low material-to-liquor ratio (1:2). High-speed drying machines have been developed to increase productivity and the machines also have a self-lubricating effect, thus, they may be run at very low maintenance cost (Teli, 1996; Lakshmanan and Raghavendran, 2017; Raichurkar and Ramachandran, 2015; Seshama et al., 2017; Teli et al., 2017; Teli and Pandit, 2017; Basak et al., 2016a).

A microwave machine is an important and useful machine for processing textile material with a low liquor ratio. Microwaves are a form of electromagnetic radiation with wavelengths between the infrared and radio frequencies, 300 MHz to 300 GHz (Basak et al., 2016a). Microwave techniques are being use in textile processing to save energy. The dyeing process depends on the dielectric property of the dye molecules and solvent. The molecular vibration or rotation during the dyeing process is controlled by the field strength of the microwave. Microwave heating takes place though ion conduction, which is a resistance type of heating. According to the field strength, the dye molecules vibrate and accelerate through the dye solution. This leads to collision of dye molecules with the surface of a fiber and provides better penetration of dye molecules inside the fiber structure. This phenomenon causes a large difference in the shade of fabric from conventional dyed fabric.

Electrochemical dyeing is another promising invention in dyeing of the textile substrates. Sulfur and vat dyes require a reducing agent in alkaline solution to make them water solubilize dyes prior to application on the textile substrate. In the reducing process, some byproducts are produced; thus, the whole solution cannot be reused for further processing. BASF is trying to reduce these vat and sulfur dyes using an electro-chimerical process (Basak et al., 2016a; Kulandainathan et al., 2007; Beaupré et al., 2006; Sala and Gutierrez-Bouzan, 2012). Although this type of electrochemical dyeing process is still at the lab scale, it may become a future green technology for the dyeing of textiles without producing much effluent. The Dyestar, GmbH and Co, and the Institute of Textile Chemistry and Textile Physics at the University of Innsbruck in Dornbirn, Austria group patented the electrochemical dyeing process, with which the Indanthrane blue E-BC dye could be reduced using electric charge. These dyes can be used several times without zero level of contamination in the dye liquor. Electrochemical dyeing is carried out by two methods, namely direct and indirect electrochemical dyeing. In the direct dyeing process, dyestuff is reduced directly between two electrodes. In the indirect method, dyestuff is not reduced directly between two electrodes but in the system a reducing agent is added to reduce the dyestuff. In the year 1993, the indirect method was patented by Thomas Bechtold. After the process, dye molecules are again oxidized in the bath and subsequently, oxidized dyestuff is reduced in the cathode portion. This process runs continuously during the dyeing process. After the dyeing process, the dyes that have not been exhausted are collected by an air oxygen treatment process. After collection of dyes, color-containing solvent, alkali, and other materials are recycled from the bath.

Various irradiation techniques, such as plasma and UV, have been found to be very useful for pretreatment of the textile followed by its dyeing. Plasma is the fourth state of matter, in which the molecules are present in an ionized state (Samanta et al., 2015a, 2006; Teli et al., 2015a, b). As system energy increases gradually, the movement of molecules and the collisions between them are also increased gradually. Therefore, the matter changes its state from solid to liquid, liquid to gas, and then gas to plasma. Plasma contains ionized and/or charged particles and radicals, but the net potential is zero; thus, it cannot be stored like other solid or liquid chemicals. Therefore, it must be generated at the time the material is to be treated. Plasma may be classified by considering the power supply, pressure, and temperature (Samanta et al., 2006). The plasma source may be categorized as either direct current (D.C.) or alternating current (A.C.), based on the electrical system used. Likewise, plasma may be classified as either atmospheric or low pressure depending upon the pressure inside the plasma reactor. Plasma may also be defined as cold or hot plasma based on the temperature of the gaseous molecules inside the plasma zone. Atmospheric pressure cold plasma is the most suitable plasma for the processing of textile materials because it allows for a continuous process, low-temperature processing, nanoscale surface engineering, water-free processing, single-step operation and shorter processing time, as well as being economical (Samanta et al., 2015a, 2010a, b, 2008, 2009, 2016; Burgos et al., 2017; Fawde et al., 2012; Devatkal et al., 2013; Patel and Patel, 2006). It has been found that highly hydrophilic cotton fabric becomes hydrophobic to superhydrophobic after application of reactive plasma irradiation of a few seconds to a few minutes. The antistatic property of polyester, antifelting of wool, hydrophilicity of synthetic fibers, improvement in dyeing/printing, improvement in biocompatibility and enhancement in adhesion strength may be accomplished by nonreactive application to reactive plasma treatment (Samanta et al., 2014a, 2015a). On the other hand, if such functionalities are to be imparted by conventional techniques, it requires a huge amount of water, chemical, energy, and long processing time, as well as the generation of effluent.

Peracetic acid bleaching and enzyme-based processing are the other innovative approaches to minimize water and the energy consumption. Mostly, sodium hypochlorite (NaOCl) and hydrogen peroxide are used in textile industries to bleach the material by redox potential. Low-temperature bleaching may be done with peracetic acid (PAA) that has more redox potential than hydrogen peroxide. Such a bleaching agent is popular for cotton, flax, and nylon fibers (Alexandra, 2009; Tenenbaum and David, 1998). In this process, the decomposition rate and consumption vary with temperature and pH of the solution bath. Enzyme-based processing of textile materials is also important for conserving energy and the environment. An enzyme, a protein catalyst prepared from cells, catalyzes only specific chemical reactions (Doshi, 2002; Shanmugasundaram and Kumaravel, 2005; Verma and Nishkam, 2003; Manickem and Prasad, 2006). Textile industries have adopted enzymatic processes for dealing with environmental issues. Textile processing, namely scouring, bleaching, softness finishing, and peroxide removal, can be carried out by enzymatic treatment. For removal of starch from the textile substrate, an amylase enzyme is used that degrades starch to lower molecular weight polymers, such as sugar, dextrin, and maltose. The scale of wool fiber is removed by proteolytic enzymes. On the other hand, catalase enzymes are used for peroxide bleaching to remove peroxide. Bio-polishing of cotton, an important process specially used in denim textiles, is practiced using cellulose enzymes. Upon treatment with such enzyme, the cotton fabric becomes smoother and softer. In spite of many advantages, enzymes are considered to work well at a specific temperature, pH level, and type of fabric (chemical composition) and the process is costlier than the conventional process. Further research effort is required in this area to formulate other enzymes that modify the other fibers.

3.3.2 Textile coloration by printing

Color can be imparted onto a textile substrate either through dyeing or a printing process. The dyeing process is a method by which the color is imparted onto the whole part of the fabric, whereas in the printing process, color is imparted onto the textile substrate in a localized area. The printing process can be categorized as pigment printing, direct printing, resist printing, and discharge printing, according to the process sequence. The conventional printing process is slow and labor intensive. Also, it requires water and a lot of energy to run the process. To address some of the issues associated with a traditional printing process, such as the water and energy requirements, digital printing has come onto the market. In digital printing, the color solution comes out from the tip of a nozzle that is controlled by a microprocessor. In the digital printing process, the fabric is printed without touching the fabric surface (Dominioni, 2003; Aston et al., 1993; Yang and Li, 2003; Hees et al., 2004; Basak et al., 2017, 2015a). Most of the time, CMYK (Cyan-Magenta-Yellow-Black) colors are used. The printing speed of a continuous ink jet printing machine is very high and it has no limitation of color. Continuous ink jet printing may be further classified into two categories, namely, continuous binary deflected ink jet and continuous multiple level deflected ink jet. Within this area, drop-on-demand technology is the latest method of printing. This is the most complex system and a lot of research is being performed in this area for further improvement.

This technique is further subcategorized depending on the thermal behavior, type of valve, and the electrostatic and acoustic nature of the ink jet. The quality of the digital printing depends on the surface tension, viscosity, pH, drying time, spreading type, and chemical nature (ecofriendly) of the ink. The printing ink could be a dye or pigment. The fabric is treated first with chemicals to ensure bleeding resistance. Sometimes curing (UV/electron beam) is required as a postprinting operation to ensure better quality. In the year 1975, Milliken launched the millitron digital printing machine, which works via the continuous valve technique to print carpet and upholstery. "True color" is the earliest commercial ink jet printing machine introduced by Stork. The DReAM machine was invented by Reggiani, Scitex vision Ltd. and Ciba specialty chemicals. This machine is suitable for bulk industrial production. Later on, a number of developments were made in this area, such as elimination of the steaming and washing steps. Recently, cotton fabric processed with the pad, dry, and cure method by using pigment-based ink jet printing has been demonstrated.

3.3.3 Spray and foam finishing of textiles

For textile finishing, the spray technique has been developed in the context of conservation of water (Uqaili and Harijan, 2011; Shaikh and Ayaz, 2009; Rathore, 2012; Alexandra, 2009; Tenenbaum and David, 1998; GG 62, 1997; Samanta et al., 2006)[1]. The process provides flexibility in terms of location-specific deposition, easy maintenance, and a high level of uniformity by uniform droplet size. Additionally, this technique requires less water, chemicals, and energy for drying and curing. Different types of functional finishes, such as antimicrobial, softness, hardness, and water repellent, can be applied by this technique. Foam finishing is another approach to reduce the water consumption and processing cost (Teli, 1996; Lakshmanan and Raghavendran, 2017; Raichurkar and Ramachandran, 2015; Seshama et al., 2017; Teli et al., 2017; Teli and Pandit, 2017; Basak et al., 2016a, 2015b). It can be generated by air blowing continuously in liquid in the presence of different chemical and foaming agents. Foam is a colloidal system containing air bubbles, which are dispersed in a liquid medium. Foam stability, bubble size, and density are the important process parameters that ultimately control the process and uniformity. Commonly, foam density varies in the range of 0.14–0.07 g/cc for foam finish, whereas it remains 0.33–0.2 g/cc for foam printing. The viscosity of the foam depends on the foam bubble density and the quantity of unfoamed liquor. The foam stability may be increased by increasing the viscosity and reducing the size of the droplets. By using this technique, dyeing, printing, and different types of functional finishes can be applied by the kiss roll or knife-over-roller coating processes.

3.3.4 Low-water and/or solvent-based coating/nanocoating of textiles

Coating is a technique by which different finishes can be imparted on the textile substrate. In the coating formulations, different types of polymers, such as poly-(vinyl chloride) (PVC), polyester, polyamide, polyurethane, poly-tetrafluoroethylene, and rubber, are used as a major component (Teli, 1996; Lakshmanan and Raghavendran, 2017; Raichurkar and Ramachandran, 2015; Seshama et al., 2017; Teli et al., 2017; Teli and Pandit, 2017; Basak et al., 2016a; Das et al., 2015). Such polymers are used to make the fabric water repellent, antimicrobial, or an electrical insulator. Sometimes pressure-sensitive adhesive can also be applied on a substrate, which is attached on the fabric first, then another layer of fabric is pressed in the presence of heat and pressure to remove water. Common adhesives used for the coating are natural rubber, styrene butadiene rubber, and acrylic polymer. Solvent-based adhesive can be used to attach microporous membrane on the textile material to impart a water repellent finish on the fabric. This type of fabric is mostly used in operation theaters in hospitals. Solvent-based polyurethane coating cures in the presence of moisture, which is sprayed on the fabric during processing, and then the fabric is passed through a nip roller. Powder coating is another technique to reduce water consumption. In this process, polymer granules of size 60–200 µm are scattered on the fabric and subsequently the granules are melted through heat treatment. Thereafter, another layer of fabric is pressed on it before the molten polymer solidifies. Such a process is fully water or solvent free. A nanocoating on the textile substrate can be applied by the plasma technique, chemical vapor deposition (CVD), or the layer-by-layer deposition technique. Such a coating does not hinder the appearance or feel of the fabric. In spite of its several advantages, this process suffers from nonuniformity that arises throughout the surface in large-scale production.

3.3.5 Use of supercritical fluid for low-water processing

Most materials have three states, solid, liquid, and gas. In the gaseous stage, molecules are randomly distributed in the system. As the temperature and pressure increase above the critical point, some fluids (gas) turn into supercritical liquid that is not purely a gas or a liquid. This type of liquid is being used in the manufacturing process to control water pollution in an economical way. In this context, carbon dioxide is very popular due to its abundant availability in nature at a low price. Apart from the price, it has other advantages, such as being nontoxic, chemically inert, recyclable, reusable, and ecofriendly, with no waste generation, low surface tension, low viscosity, and ease of handling (Teli, 1996; Lakshmanan and Raghavendran, 2017; Raichurkar and Ramachandran, 2015; Seshama et al., 2017; Teli et al., 2017; Teli and Pandit, 2017; Basak et al., 2016a, 2015c). In a supercritical fluid-based dyeing process, fabric is dyed after it is wrapped on a perforated stainless-steel tube inside an autoclave. Dyestuff is placed at the bottom of the chamber and then the temperature is raised with continuous purging of carbon dioxide with constant staring. After reaching the critical temperature, the pressure is raised and subsequently the pressure and temperature are maintained for 60 min for the dyeing cycle. After the process, the pressure is released and the carbon dioxide and residual dyestuff are recycled. The dyed dry sample is washed with acetone to remove excess dyestuff from the fabric surface. Because the dyeing process takes place in a virtual gas phase, a high degree of levelness is achieved as compared to conventionally dyed fabric.

3.3.6 Ultrasonic wave-assisted textile processing

According to sound wave theory, sound may be classified as infrasound with frequency below 16 Hz, audible sound in a frequency range of 16–20 kHz, and ultrasound with frequency greater than 20 kHz. The nature of ultrasound is similar to audible sound with the only difference being the wavelength. Such a wave cannot be recognized by the human body. Transverse and longitudinal waves are transmitted through solid media but in liquid rarefaction and compression are generated by the longitudinal vibration of molecules. This causes the formation of bubbles of 500 nm that can collapse and create a shock wave in the system. The formation of such vapor bubbles depends on the intensity and frequency of the wave, the temperature, and the vapor pressure of the liquid. High-velocity microjets form at a speed of 100–160 m/s toward the surface when the bubbles collapse. This method is being used in various textile processes, namely, desizing, scouring, bleaching, mercerization, and auxiliary processes (Rahman et al., 2013; Basak and Ali, 2017; Basak et al., 2014; Smith and Thakore, 1991; Rathi et al., 1997; Yachmenev et al., 1998). Desizing using ultrasonic techniques is economical as compared with conventional processes. It has also been found that oil is removed completely from ultrasound-treated cotton and nylon fabrics, even though the untreated samples contained oils. The stability and the fiber properties are unaffected by this process. The bleaching (peroxide) rate and whiteness index of the fabric after bleaching improve significantly through ultrasound (20 kHz) treatment. Additionally, this process helps in saving energy by reducing the processing temperature. Ultrasound-treated fabric shows better absorbency and fastness as compared with conventionally treated fabric. Sometimes, in the mercerization process, this technique is used to speed up the process 2–3 times for 100% cotton fabric.

Dyeing using the ultrasonic technique was started in the year 1941. Cotton fiber with direct dye, wool fiber with acid dye, and polyamide and acetate fibers with dispersed dyes are used for coloring textile materials. When an ultrasonic wave passes inside the dye solution, cavitation formation takes place. This process enhances the dyeing rate of polyamine and acetate dyeing with dispersed dye. This technique is more efficient for insoluble nonpolar dye and for hydrophobic fibers, such as polyester and nylon. The uniformity of color on the fabric can be improved by ultrasonic treatment. The frequency of ultrasound plays a major role in the efficiency of textile coloration. The efficiency of ultrasound with frequency of 50–100 c/s is negligible, but significant efficiency has been achieved with ultrasound frequency of 22–175 c/s. A uniform dispersion of dye molecules from clusters in solution is achieved with the ultrasound technique. Therefore, better uniformity of color is obtained after the dyeing process. Another explanation is more popular, that is, the degassing technique in which trapped gas molecules are removed from the solution bath and enter fiber crossover regions. Such a process enhances the probability of interactions between dye molecules and the fiber surface; thus, the dyeing rate and efficiency improve significantly as compared with the conventional dyeing process. Another attribute of the ultrasound treatment during dyeing may be explained by a diffusion mechanism that leads to mechanical action and heating for better diffusion of dye molecules inside the fiber. However, in the case of pigment, it helps in better dispersion in the solution. This additional advantage of mechanical agitation is not available in conventional processing. The ultrasonic system contains major components, such as the generator and converter. Usually, the generator converts an electrical signal of 50–60 Hz to electrical energy of high frequency that further converts into mechanical vibration in the liquid medium through the transducer. The transducer vibrates longitudinally in the liquid. Cavitation formation takes place inside the liquid as the wave propagates. A portable machine has been designed for the dyeing of yarn and fabric.

3.4 Textile processing using different irradiation techniques

Ultraviolet (UV) and plasma irradiation have been introduced for saving water, energy, and chemicals in textile processing, which is otherwise a water- and energy-intensive process. Actually, these rays/ionized gas are composed of ions, charged particles, neutrons, electrons, and excited species. When the material surface is irradiated, it leads to physicochemical changes of the surface at the nanometer level. As a result of this, dyeing and the other wet processing of the material become easier. For example, wool fabric was pretreated with strong UVC irradiation (172, 212 nm) in different atmospheres of oxygen, nitrogen, and air. It was observed that the wool fabric irradiated for 15 min in a nitrogen atmosphere showed 100% exhaustion of acid dye as compared to the 73% for the untreated samples after 20 min dyeing time. Treated wool fabric showed saturation dye uptake as well as an improved rate of dyeing. This is due to the fact that the UV treatment in nitrogen atmosphere generated more amine groups on the wool surface at the nanometer level and these act as extra dye sites for anionic acid dye molecules. It has also been reported that UV rays can etch the wool surface, generate free radicals, distort the surface scales, and assist in dye uptake. In contrast, wool fabric UV-irradiated in air and oxygen atmosphere showed an uptake of cationic basic dye molecules in less dyeing time due to the formation of anionic carboxylic and hydroxyl groups on the fabric surface. Overall, it can be said that the UV irradiation helps to shorten the wool dyeing process by lowering the dyeing temperature and time, thus saving the considerable amount of energy that would otherwise be required in conventional dyeing (Gupta and Basak, 2010; Basak et al., 2016b). UV and plasma pretreatment have also been attempted for the removal of scales from the surface of wool fiber (antishrink finish), as an alternative to the use of harmful chlorinated chemicals. Like UV pretreatment, plasma technology also helps to reduce the dyeing cycle and increase the rate of dye uptake. Additionally, plasma has been used to make cotton fabric hydrophobic by using plasma-assisted in situ/post polymerization of hydrocarbon or fluorocarbon precursors (Samanta et al., 2016). However, the major disadvantages of this irradiation process are the uniformity in dyeing and other functional finishing. Additionally, the fastness of dyed samples is always unsatisfactory, as most of the dye molecules get deposited/reacted on the outer layer of the substrate. Laser rays and gamma rays have also been validated for textile processing; however, these techniques are not popular or successful due to the high capital cost and the uniformity of the treatment.

3.5 Effluent generation and treatment

The textile industry consumes a large quantity of water, dyes, pigments, soap, detergent, various chemicals and auxiliaries, steam, and electrical energy, that lead to generation of effluent with high BOD and COD loads, suspended solids, mineral oils, and residual dye (Saxena et al., 2017). These chemicals and other environmentally unfriendly residues present in the processed water may lead to air and water pollution. Indeed, increasing the pollution from waterborne sources, including the textile industry, results in adverse health effects in humans. The sources of water pollution are mainly process wastes from processes, such as sizing, desizing, scouring, bleaching, mercerizing, dyeing, printing, and finishing. Additionally, the contaminated water from the rinsing and washing after each processing step adds with the quantity of effluent already generated directly from the processing. In the dyeing of textile, there are a number of ingredients for which wastewater is generated, as shown in detail in Table 3. 5.

3.5.1 Different techniques to reduce pollution load

There are a number of suitable approaches that can be utilized so as to minimize the effluents or pollution load generated by the textile industry. A few of the approaches in this direction are to minimize/reuse the raw materials, the adoption of ecofriendly

Table 3. 5 Amounts of textile wastewater generated by different departments (GG 62, 1997; Saxena et al., 2017; Shukla, 2007)

Process unit in textile industry	pH value	Suspended solids (mg/L)	BOD (mg/L) (5 days at 20°C)	COD (mg/L)
Desizing (starch, CMC, PVA resins, fats, and waxes)	8.6–10	2290–2670	1000–1080	1650–1750
Kiering	10.9–11.8	1960–2080	2500–3480	12800–19600
Bleaching (hypochlorite, chlorine, hydrogen peroxide, acids)	8.4–10.9	200–340	87.5–535	1350–1575
Mercerizing (alkali, etc.)	8.1–9.2	160–430	100–1222	246–381
Dyeing (dyestuffs, mordents and reducing agents, such as sulfides, hydrosulfite, salt, caustic soda, acetic acid, and soap)	9.2–11	360–370	130–820	465–1400
Printing (dyes, starch, gums, oil, china clay mordents, acids, and metallic salts)	6.7–8.2	250–390	135–1380	410–4270

chemicals/processes, and, if pollution is unavoidable, then to eliminate or reduce its aftereffects. Reduction in the concentration/amount of chemicals used by optimizing their usage so as to reduce their adverse effects is necessary; however, it is not always possible in the textile processing industry (GG 62, 1997; Saxena et al., 2017; Shukla, 2007). Other suitable approaches are discussed in detail below.

Reducing the wastewater volume is important for minimizing the quantity of the effluent and can be achieved by the following methods: reducing the number of washing steps, the use of hot water for washing, the use of low-liquor-ratio systems. Wastewater can also be reduced by the recovery and reuse of the chemicals discharged from different parts of the process, for example, recovery of the sizing agents from the desizing bath or wool grease from the wool scouring bath. Along these lines, the recycling of such textile printing chemicals as polysaccharide thickeners, namely, sodium alginate (A), carboxy methylated guar gum (CMG), and carboxy methylated cellulose (CMC) have been carried out. This kind of polysaccharide is widely used as a thickener in textile printing to ensure the desired viscosity of the print paste (Samanta et al., 2015b). Likewise, polyvinyl alcohol (PVA), another sizing chemical used for sizing polyester/cotton blended yarns, may be recovered by the ultrafiltration method. Additionally, recovery of caustic soda from the processed mercerizing water is also an important part of the waste minimization.

A reduction in the concentration of waste is also possible by chemical substitution, that is, by the use of ecofriendly chemicals instead of hazardous chemicals. As an example, it researchers have suggested using CMC and Poly(vinyl alcohol) in place of starch and gelatin, used in the sizing of cotton warp and other textile yarns. Use of mineral acids, synthetic detergent, and ecofriendly stabilizers in place of acetic acid, soap, etc. is another important solution, as the mineral acids have zero BOD and 60% the COD of acetic acid. As per literature, synthetic detergent has a BOD level of 0%–22%, whereas soap increased the BOD level by 140%. As far as the dyeing process is concerned, it is beneficial to use reactive dye as much as possible instead of azo, sulfur, vat, and acid dyes. Concerning the use of reactive dyes, multifunctional reactive dye is beneficial for better fixation and to ensure minimum waste in the textile industry. Other possibilities for minimizing the waste effluent include using different enzymes and exploring ecofriendly crosslinkers (soya resin, butane-tetra-carboxylic-acid, citric acid, etc.,) instead of formaldehyde-based crosslinking agents (Di-methyl-dihydroxy-ethylene-urea, melamine formaldehyde, etc.) (Samanta et al., 2014b). Waste minimization may also be possible through process modification and is achieved by new processes, such as foam finishing, transfer printing, and digital printing. Now a day's worth of reduction in the water consumption is an important need along with reducing the energy requirement. Recently, advanced methods have been introduced for effluent treatment based on adsorption, ion exchange, and membrane filtration. It can be done in three ways, such as reverse osmosis, ultrafiltration and nanofiltration. Other methods used for the effluent treatment are ozonization, evaporation, crystallization, and many more.

3.6 Conclusion and the future prospects

Textile chemical processing is necessary to ensure requisite value addition to it. In the necessary process steps, starting from preprocessing, dyeing and, functional finishing, the textile industry consumes a significant amount of water and various chemicals as discussed in the earlier sections. At the same time, a large quantity of water-based effluent is generated, which is discharged into the water stream contaminated with residual dyes, pigments, and other chemicals. Some of the chemicals used in preparatory processing, dyeing, and finishing cause water, air, and soil pollution, resulting in an adverse effect on aquaculture and agriculture. In the last few decades, industries and the researchers have adopted different modern approaches for the treatment of discharged effluent. A number of innovative processes have also been adopted to reduce the required quantity of water. In the recent past, different dyeing and pretreatment machines (jet dyeing, beck dyeing, jigger, continuous process, such as CBR, CDR, etc.) have been commercialized, which can run at lower material-to-liquor ratio (within 1:5). In addition, the production level is also higher in these machines compared with conventional dyeing and pretreatment machines. In the recent past, due to more awareness of personal health and hygiene and a desire to preserve natural resources, natural fiber-based textiles that are dyed and finished with natural ingredients, such as coloration with natural dyes, bio-polishing with enzymes, antimicrobial finishing with bio-molecules/biomaterials, UV protective finishing with natural dyes, and flame retardant finishing with bio-macromolecules and plant extract, have been gaining importance in the commercial market. Efforts are also ongoing to slowly replace a portion of the environmentally unfriendly synthetic chemicals and auxiliaries with natural extracts, bio-molecules, biomaterials, and biopolymer. Water-free plasma, UV, laser, ozone, sand blast, digital printing, and similar technologies have been used as a pretreatment, posttreatment, for in situ processing or postpolymerization, and for surface activation, oxidation, etching, polymerization, printing, and coating textiles to impart various functionalities from the point of view of water, energy, chemical, cost and, environmental pollution. However, there are several challenges, including the uniformity, the fastness properties, the stability of the finish to aging, and the initial and operational costs. Nanotechnology-based processing has the advantage of lowering the chemical requirements in a particular process. However, in most cases, it has been found that the treated fabric exhibits lower washing and rubbing fastness and the application is not uniform. The supercritical carbon dioxide fluid-based treatment ensures good levelness; however, the process is costly and restricted only to the domain of synthetic fiber processing. Low-liquor foam finishing is another upcoming area; however, it is not used much due to the uniformity challenge along with the difficulty in maintaining the proper viscosity, bubble size, etc. Digital printing is very popular nowadays and it provides a uniform clear print on the textile substrate in very little time. Concerning the effluent treatment, different modern techniques, such as adsorption, ion exchange, membrane filtration, reverse osmosis, ultrafiltration, and nanofiltration, are very useful in the commercial market. Other methods for effluent treatment, such as ozonation, evaporation, and crystallization, are also important for commercial applications.

References

Ahmed, K., Das, M., Islam, M.M., Akter, M.M., Islam, M.S., Al-Mansur, M.A., 2011. Physicochemical properties of tannery and textile effluents and surface water of River Buriganga and Karnatoli, Bangladesh. World Appl. Sci. J. 12, 152–159.

Alexandra, A., 2009. Yet another' footprint'to worry about: water. Wall Street J. 17, 24–29.

Aston, S.O., Masselink, H., Provost, J.R., 1993. Jet Printing With Reactive Dyes. Soc. Dyers Color. 109, 147–152.

Basak, S., Ali, S.W., 2017. Leveraging flame retardant efficacy of the pomegranate rind extract, a novel biomolecule, on lingo-cellulosic materials. Polym. Degrad. Stab. 144, 83–89.

Basak, S., Samanta, K.K., Chattopadhyay, S.K., Das, S., Bhowmik, M., Narkar, R., 2014. Fire retardant and mosquito repellent jute fabric treated with thio-urea. J. Text. Assoc. 74, 273–280.

Basak, S., Samanta, K.K., Saxena, S., Chattopadhyay, S.K., Mahangade, R., 2015a. Flame retardant cellulosic textile using banana pseudostem sap. Int. J. Cloth. Sci. Tech. 27, 247–258.

Basak, S., Samanta, K.K., Chattopadhyay, S.K., Narkar, R., 2015b. Self-extinguishable lingocellulosic fabric by using banana pseudostem sap. Curr. Sci. 108, 372–380.

Basak, S., Samanta, K.K., Chattopadhyay, S.K., Narkar, R., 2015c. Thermally stable cellulosic paper made using banana pseudostem sap, a wasted byproduct. Cellul. 22, 2767–2773.

Basak, S., Samanta, K.K., Chattopadhyay, S.K., Pandit, P., Maiti, S., 2016a. Green fire retardant finishing and combined dyeing of proteinous wool fabric. Color. Technol. 132, 135–143.

Basak, S., Patil, P.G., Shaikh, A.J., Samanta, K.K., 2016b. Green coconut shell extract and boric acid: new formulation for making thermally stable cellulosic paper. J. Chem. Technol. Biotechnol. 91, 2871–2890.

Basak, S., Samanta, K.K., Saxena, S., Chattopadhyay, S.K., Parmar, M.S., 2017. Self-extinguishable cellulosic textile from Spinacia Oleracea. Ind. J. Fibre Text. Res. 42, 215–222.

Beaupré, S., Dumas, J., Leclerc, M., 2006. Toward the development of new textile/plastic electrochromic cells using triphenylamine-based copolymers. Chem. Mater. 18, 4011–4018.

Burgos, E.C.R., Hernender, A.B., Artiaga, L.N., Kacanlova, M., Garcia, F.H., Lopez, J.L.C., Barrachiana, L.L.C., 2017. Antimicrobial activity of the pomegranate peel extracts as effected by cultivation. J. Sci. Food Agric. 97, 525–532.

Das, S., Bhowmick, M., Chattopadhyay, S.K., Basak, S., 2015. Application of biomimicry in textiles. Curr. Sci. 109, 893–901.

Devatkal, S.K., Jaiswal, P., Jha, S.N., Bhardawaj, R., Viswas, K.N., 2013. Antibacterial activity of the aqueous extract of the pomegranate peel against pseudomonous stutzeri isolated from poultry meat. J. Food. Sci. Technol. 50, 555–560.

Dominioni, A.C., 2003. European Textile Ink Jet Conference. Basel, Switzerland.

Doshi, R., 2002. Impact of bio-technology on the textile industry. Asian Textile J.

Fawde, O.A., Makunga, N.P., Opara, U.L., 2012. Antibacterial antioxidant and tyrosinase inhibition activities of pomegranate fruit peel methanolic extract. BMC Complement. Altern. Med. 12, 200–210.

GG 62, 1997. Water and chemical use in the textile dyeing and finishing industry.Guide Produced by the Environmental Technology Best Practice Programme, UK. Retrieved from: http://www.wrap.org.uk/sites/files/wrap/GG062.pdf. (Accessed 5 May 2016).

Gómez, N., Sierra, M.V., Cortelezzi, M.V., Capítulo, A.R., 2008. Effects of discharges from the textile industry on the biotic integrity of benthic assemblages. Ecotoxicol. Environ. Saf. 69, 472–479.

Gupta, D., Basak, S., 2010. Surface functionalisation of wool using 172nm UV Excimer lamp. J. Appl. Polym. Sci. 117, 3448–3453.

Hees, U., Freche, M., Kluge, M., Provost, J., Weiser, J., 2004. Ink jet interactions in ink jet printing-the role of pre-treatments.Textile Ink Jet Printing-A Review Of Ink Jet Printing Of Textiles, Society of Dyers and Colourists Technical Monograph, 7, pp. 44–56.

Kulandainathan, M., Patil, K., Muthukumaran, A., Chavan, R.B., 2007. Review of the process development aspects of electrochemical dyeing: Its impact and commercial applications. Color. Technol. 123, 143–151.

Lakshmanan, S.O., Raghavendran, G., 2017. Low water-consumption technologies for textile production. Sust. Fibr. Text. 7, 243–265.

Lloyd, B.J., Helmer, R., 1991. Surveillance of Drinking Water Quality in Rural Areas. Longman Scientific and Technical. Wiley, New York, pp. 34–56.

Manickem, M.M., Prasad, J.G., 2006. Application of biotechnology in textile. Asian Dyer 67, 34–49.

Mansour, H.B., Houas, I., Montassar, F., Ghedira, K., Barillier, D., Mosrati, R., Chekir-Ghedira, L., 2012. Alteration of in vitro and acute in vivo toxicity of textile dyeing wastewater after chemical and biological remediation. Environ. Sci. Pollut. Res. 19, 2634–2643.

Patel, B.H., Patel, R.N., 2006. Plasma aided wet chemical treatment: an update. Ind. Text. J. 31-34.

Rahman, N.A.A., Tajuddin, R., Tumin, S.M., 2013. Optimization of natural dyeing using ultrasonic method and biomordant. Int. J. Chem. Eng. Appl. 4, 205–210.

Raichurkar, P., Ramachandran, M., 2015. Effluent generated from the textile process industries. Int. J. Text. Eng. Proc. 1, 47–50.

Rathi, H.N., Mock, N.G., McCall, E.R., Grady, L.P., 1997. Ultrasound aided open width washing of mercerized 100% cotton twill fabric. AATCC, 254–262.

Rathore, J., 2012. Studies on pollution load induced by dyeing and printing units in River Bandi at Pali, Rajasthan, India. Int. J. Environ. Sci. 3, 735–742.

Sala, M., Gutierrez-Bouzan, M.C., 2012. Electrochemical techniques in textile processes and wastewater treatment. Int. J. Photoenergy 2012, 1–12.

Samanta, K.K., Jassel, M., Agrawal, A.K., 2006. Atmospheric pressure glow discharge plasma and its applications in textile. Ind. J. Fiber Textile Res. 31, 83–98.

Samanta, K.K., Jassel, M., Agrawal, A.K., 2008. Formation of nano-sized channels on polymeric substrates using atmospheric pressure glow discharge cold plasma. J. Nanotechnol. Appl. 4, 71–75.

Samanta, K.K., Jassel, M., Agrawal, A.K., 2009. Improvement in water and oil absorbancy of the textile substrate by atmospheric pressure cold plasma treatment. Surf. Coat. Technol. 203, 1336–1342.

Samanta, K.K., Jassel, M., Agrawal, A.K., 2010a. Antistatic effect of atmospheric pressure glow discharge cold plasma treatment on textile substrates. Fibers Polym. 11, 431–437.

Samanta, K.K., Jassel, M., Agrawal, A.K., 2010b. Atmospheric pressure plasma polymerizationof 1,3-butadiene for hydrophobic finishing of textile substrates. J. Phys. Conf. Ser. 208, 012098.

Samanta, K.K., Basak, S., Chattopadhyay, S.K., 2014a. Environment-friendly textile processing using plasma and UV treatment. In: Roadmap to Sustain. Text. Cloth. Springer, India, pp. 161–201.

Samanta, K.K., Basak, S., Chattopadhyay, S.K., 2014b. Eco-friendly coloration and functionalization of textile using plant extracts. In: Roadmap to Sustain. Text. Cloth. Springer, India, pp. 263–287.

Samanta, K.K., Basak, S., Chattopadhyay, S.K., Gayatri, T.N., 2015a. Water-free plasma processing and finishing of apparel textiles. In: Hand B. Sustain. Appar. Prod. CRC Press, Boca Raton, Fl, pp. 3–37.

Samanta, K.K., Basak, S., Chattopadhyay, S.K., 2015b. Recycled fibrous and nonfibrous biomass for value-added textile and nontextile applications. In: Environ. Implic. Recycl. Recycl. Prod. Springer, India, pp. 167–212.

Samanta, K.K., Gayatri, T.N., Saxena, S., Basak, S., Chattopadhyay, S.K., Arputharaj, A., Prasad, V., 2016. Hydrophobic functionalization of cellulosic substrates using atmospheric pressure plasma. Cellul. Chem. Technol. 50, 745–754.

Saxena, S., Raja, A.S.M., Arputharaj, A., 2017. Challenges in sustainable wet processing of textiles. In: Textiles and Clothing Sustainability. Springer, Singapore, pp. 43–79.

Seshama, M., Khatri, H., Suther, M., Basak, S., Ali, S.W., 2017. Bulk Vs nano ZnO: influence of fire retardant behaviour on sisal fibre yarn. Carbohydr. Polym. 175, 256–262.

Shaikh, H., Ayaz, A.M., 2009. Water conservation in textile industry. Pak. Textile J. 6, 48–51.

Shanmugasundaram, O.L., Kumaravel, S., 2005. Role of enzymes in textile processing Industry. Asian Dyer.

Shukla, S.R., 2007. Pollution abatement and waste minimization in textile dyeing. Environ Asp. Text Dyeing 116–148.

Smith, C.B., Thakore, A.K., 1991. Study of the influence of ultrasound on enzymatic treatment of cotton fabric. Textile Color. Chem. Am. Dyestuff Rep. 23, 23–25.

Solmaz, S.K.A., Birgül, A., Üstün, G.E., Yonar, T., 2006. Colour and COD removal from textile effluent by coagulation and advanced oxidation processes. Color. Technol. 122, 102–109.

Teli, M.D., 1996. Novel developments in dyeing process control. Ind. J. Fiber Textile Res. 21, 41–49.

Teli, M.D., Pandit, P., 2017. Novel method of ecofriendly single bath dyeing and functional finishing of wool protein with coconut shell extract biomolecules. ACS Sustain. Chem. Eng. 5, 8323–8333.

Teli, M.D., Pandit, P., Samanta, K.K., 2015a. Application of atmospheric pressure plasma technology on textile. J. Text. Assoc. 75, 422–428.

Teli, M.D., Samanta, K.K., Pandit, P., Basak, S., Chattopadhyay, S.K., 2015b. Low-temperature dyeing of silk fabric using atmospheric pressure helium/nitrogen plasma. Fibers Polym. 16, 2375–2383.

Teli, M.D., Pandit, P., Basak, S., 2017. Coconut shell extract imparting multifunctional properties to lingo-cellulosic material. J. Ind. Text. 24, 98–105.

Tenenbaum, A., David, J., 1998. Tackling the Big Three Environmental Health Perspective,106, A234.

Uqaili, M.A., Harijan, K. (Eds.), 2011. Energy, Environment and Sustainable Development. Springer Science and Business Media, New York.

Verma, N., Nishkam, A., 2003. Enzymes in wet processing. Ind. Text. J. 37, 49–56.

World Bank, 2010. In: A detailed analysis on industrial pollution in Bangladesh.Workshop Discussion Paper Transfer, EPA-625/3-74-004, World Bank Dhaka Office, Dhaka, Bangladesh.

Yachmenev, V.G., Blanchard, E.J., Lambert, A.H., 1998. 481. AATCC, 427.

Yang, Y., Li, S., 2003. Cotton fabric inkjet printing with acid dyes. Text. Res. J. 73, 809–814.

Water withdrawal and conservation—Global scenario

P. Senthil Kumar, C. Femina Carolin
Department of Chemical Engineering, SSN College of Engineering, Chennai, India

4.1 Introduction

Ninety percent of worldwide energy creation is water intensive, putting a huge demand on water assets for the conversion process (Jimenez et al., 2015). The major natural resource need on the Earth is water. Individuals utilize water to develop crops and create a wide assortment of items. Assemblages of water also support an extensive variety of environmental needs. As indicated by the World Bank, 21% of transferable illnesses in India are related to risky water, and as indicated by the WHO, 97 million Indians need access to safe water today (Srikanth, 2009). The business and energy segments of India also use water wastefully. Recently, the Ministry of Environment and Forests demonstrated that the aggregate sum, gathered as water tax from ventures in various states, has expanded by Rs.40 crores in the previous 3 years. For single product generation of items, such as denim, cotton, and wool, industry uses a wide range of water. The textile industry is the same as a chemical industry and causes pollution of the same type. The textile industry is the second most harmful industry in the world because most of the chemicals and dyeing materials are widely used in the textile and fashion industries, which leads to environmental problems. It uses water in various preparation tasks. In the textile industry, particularly in the manufacturing process, next to chemicals, the water consumption and usage is high. For example, in the washing process, a huge amount of water is used to remove excess dyes from garments. Especially in wet processing, the textile industries consume water extensively. Likewise, most texture arrangement steps, including desizing, scouring, bleaching, and mercerizing, utilize water. The water used fluctuates, generally based on the industry, the particular procedures used, the hardware, etc. Material activities change incredibly in water utilization. Because the textile industry is highly water intensive, regions of water scarcity have developed. Some researchers have suggested that water scarcity can lead to the depletion of the ecosystem and hence the balance between the supply and demand for water needs to be analyzed (Gleeson et al., 2012). High water consumption can lead to water becoming inaccessible, which means it cannot be used by other regions (Lee et al., 2018). There are three noteworthy reasons for water utilization: evaporation, incorporation into products, and degradation to a quality not fit for future use.

A large portion of the water utilized as a part of the assembly process comes back to nature somewhere else; the demand influences nearby accessibility for different clients. A lot of wastewater is produced in the coloring and fading forms, bringing about contaminated effluents that regularly wind up in water bodies. Water used by industry is withdrawn from the place due to which water demand gets started. High water use also leads to the generation of wastewater with negative impacts to the environment and human health. For various reasons, the textile industry is compelled to follow guidelines based on water consumption and water withdrawal. The issues related to the wastewater are depicted in Fig. 4.1. Consequently, it is possible to decrease water use through evaluating and recognizing the causes of waste. It is easy to save up to 40%–50% of the water by suitable remedial measures, such as proper control, quality control, and treatment methods. Various methods are followed in the textile and fashion industries to avoid the overconsumption and withdrawal of water. Techniques, such as counter-current washing, reuse of soapy wastewater, use of flow control valves, good housekeeping, and use of automatic shutoff valves, are widely-used techniques in the textile industry. Based on the consumption of water in the textile industry, the processes are divided into two types. Table 4.1 presents the wet and dry processes that are followed. This chapter aims to give a comprehensive

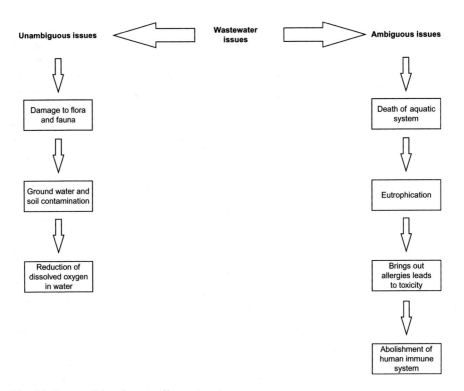

Fig. 4.1 Issues arising from textile wastewater.
Based on Verma, A.K., Dash, R.R., Bhunia, P., 2012. A review on chemical coagulation/flocculation technologies for removal of colour from textile wastewaters. J. Environ. Manage. 93, 154–168.

Table 4.1 Wet and dry processes in the textile industry

S.no	Wet process	Dry process
1	Scouring process	Picking
2	Desizing	Spinning
3	Bleaching	Weaving
4	Milling	Heat setting
5	Dyeing	Singeing

Based on Sharma, D.K., 2015. Water and wastewater quantification in a cotton textile industry. IJISET 2, 288–299.

analysis of water withdrawal and consumption in the textile industry. We also illustrate the laws and standards of water withdrawal and consumption in the textile industry. Additionally, we also describe technical and policy suggestions for saving water in the world.

4.2 Water withdrawal and conservation—a global scenario

Our Earth's freshwater assets are limited, including just 2.5% of the aggregate freshwater accessible from both surface water and the icecaps. India and China are the two largest cloth producers in the world and the industry contributes the highest output to their countries (Lin and Moubarak, 2013). The textile industry is one of the major industries in the world economy (Ranganathan et al., 2007). While the textile and fashion industries use monstrous quantities of water in the generation of products, perhaps a bigger issue is the creation of water contamination, especially in developing countries. An absence of stringent controls on companies in numerous nations implies that material makers habitually dump wastewater straight into the environment. This pollution has been connected with expanded rates of growth in asthma, and laborers have encountered second and severe burns due to the use of chemicals. Such effects are also found in countries, for example, Indonesia, China, and India, where the regulations are not strict. The textile industry is reliant on water for all means of manufacturing. Roughly 500 gal of water are used to create only one set of pants and 425,000,000 gal of water are used every day in the fashion industries in India. The major cause of water pollution is due to the textile industry and additionally it is the most water intensive of industries. In China, which is the greatest material exporter in the world, water use is of about 59 billion m^3 and 2 million tons in 2014 (World Trade Organization, 2016). For the material business in the European Union, which is the second material exporter in the world (World Trade Organization, 2016), yearly water use is 600 million m^3 (Vajnhandl and Valh, 2014). As a rising economy, Turkey is at present seeing a rapid industrial improvement and related extreme asset consumption. Water demand is increasing here due to the development of various industries. The "Turkey Water Report" says that the water utilization in industries between 1990 and 2008 increased 50.2% from 30.6 to 46.0 billion m^3 (Alkaya and Demirer, 2015).

Projections show that between 2008 and 2030 water utilization will increase almost three-fold to 112.0 billion m^3. During the same period (2008–30), industrial water utilization is projected to increase tremendously from 5 to 22 billion m^3. As it were, the share of the industrial water utilization will be required to increase from 10.9 to 19.6%. For the material business in Turkey, which has quickly created and represents around 35% of general Turkish fare limit, water utilization demonstrates a substantial scope of qualities from 50 to 100 m^3/ton of completed material (Kiran-Ciliz, 2003). A large portion of the water utilized as a part of the manufacturing comes back to nature somewhere else; the request influences neighborhood accessibility for different clients. By 2030, the water withdrawal is expected to increase from 16% to 22% and this will start initially in China. Water consumption in Europe is about 40% because it is one of the largest water users in the manufacturing industries (Vajnhandl and Valh, 2014).

A large quantity of good-quality water is required for textile operations as water is utilized as a medium and dissolvable for the disintegration of process chemicals and additionally as a washing-off gas, other than steam for warming the bath. The amount of water used for handling depends on the kind of material being prepared, the nature of the colors, and the type of handling apparatus; 50–100 L water (world normal) is needed for preparing 1 kg of material. In the scouring and bleaching processes, normal fibers require a larger amount of water than synthetic fibers because of the high impurity level of normal fibers. New advances open up the possibility of performing these activities without using water as a medium so that the contamination issues may be avoided. The use of supercritical liquids as a handling medium and the use of plasma are two such rising innovations discussed in the following sections, however, the innovation of using a laser to accomplish a blurred watch and exhausted impact on denim materials is another waterless innovation. A changing level of shading expulsion with practically zero harm to the other properties of denim material can be accomplished by using diverse laser parameters. As it is a waterless procedure and has a high potential for development, it has an edge over other regular methods of preparation (Dascalu et al., 2000; Ondogan et al., 2005). The water consumption in textile industries can be reduced by various methods, such as:

1. Plasma dyeing
2. Supercritical carbon-dioxide dyeing method

4.2.1 Plasma dyeing

Plasma is created when a substance in its vaporous state absorbs energy and gets ionized into charged atoms and free electrons. It is comprised of positive and negative particles, electrons, neutrals, radicals, and photons. It can react with a larger number of substances. Existing substance bonds are broken and new bonds are formed all the while, subsequently presenting new functionalities. Also, some physical changes may likewise happen. These physicochemical changes occur at the surface and the mass remains largely unaffected. Utilization of plasma is ecofriendly for the preparation of textile material. It has numerous applications in textile materials in which the surface properties determine different end-use properties of the material, for example, wettability, dyeability in the case of wool, and water protection. It may be possible to impart surface attributes to meet particular prerequisites by proper determination of the plasma composition, that is, the choice of gas (with at least one from O_2, N_2, H_2, air, Ar, He, NH_3, hydrocarbons, and fluorocarbons), and also the process conditions, for example, treatment time, power, weight, and gas stream rate. Plasma can be utilized to aid in the expulsion of contaminants, completing and estimating operators from the textile materials. PVA is fundamentally utilized for slicing engineered yarns and might be utilized as an optional estimating specialist for starch to the cotton yarns. A total PVA measure expulsion is difficult and requires high energy and water utilization as it is dissolvable in boiling water. Environmental plasma treatment has been found to increase the cold-water solvency of PVA on cotton (Cai et al., 2003). Plasma treatment has been used by the authors to reduce the time during cotton souring to remove starch from cotton (Kan et al., 2014; Sun and Stylios, 2004). Plasma treatment has been utilized for making fibers, for example, PET and PP hydrophilic (Mehmood et al., 2014). Oxygen plasma is normally appropriate for such applications. It can decrease the wetting time of these substrates and improve their antistatic and bond properties. Surface scratching created by plasma treatment dramatically improves the inkjet printability of PP. Another principle use of plasma in textile materials has been for creating water-, stain-, and oil-repellent properties. These highlights are accomplished by the use of plasma containing fluorine atoms. The upside of the process is that there is no adjustment in the appropriate properties of materials, for example, air penetrability due to the nanoscale of the operation and the operation is finished in a single step. No further drying and curing is required; consequently, there is a reduction in the use of water, time, and chemicals.

Plasma treatment can improve the capacity of polyester, fleece, and polypropylene strands to hold dampness on their surface. Cotton texture treated with hexamethyldisiloxane gas shows a smooth surface and an expanded contact edge of the water. It can enhance the bond property of substance covering and increase the color partiality of textile products. Oxygen plasma can be utilized for completing the antifelting of fleece texture. Plasma gas may be utilized to give UV resistance, an antibacterial quality, waterproofing, fire protection, and so forth. It reduces the preparation span and warmth vitality in coloring forms by expanding the hydrophilicity of specific materials, for example, hemp, wool, and polypropylene. Inkjet printing is a technique for printing materials having a hydrophilic property. Polypropylene materials are difficult to color and print as a result of their hydrophobic nature. Plasma pretreated polypropylene material has great shading speed and results in better adhesion with inkjet printing. Plasma innovation is utilized for surface treatment and is advantageous compared with the regular process because it does not modify the innate properties of the materials. It is done at the dry stage, and thus no cost is associated with treating effluent. There are various techniques used to set up the ionization of gases, including glow discharge, corona discharge, and dielectric barrier discharge. Plasma innovation reduces the handling cost associated with preliminary procedures by constraining the utilization of chemicals and energy and diminishing the contamination stack. Fig. 4.2 explains the application of plasma technology in textile processing.

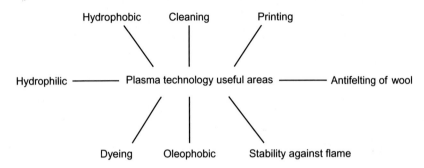

Fig. 4.2 Application areas of plasma technology in textile processing.

4.2.2 Supercritical carbon dioxide dyeing method

Supercritical carbon dioxide is utilized as a coloring medium to color manufactured filaments and textures. Extensive research studies have been performed as a result of natural issues and the increase in the cost of water in typical coloring forms (De Giorgi Cadoni et al., 2000; Youssef, 2000). Carbon dioxide is the best choice for delivering supercritical liquids because it is nontoxic and nonflammable, it is used in the nourishment and refreshment industry, and it is generated in large quantities from burning procedures. A gas is a supercritical liquid at a temperature and pressure above its critical point, at which distinct gas and fluid stages do not exist. Supercritical fluids can spread out along a surface more easily than a true fluid because they have much lower surface pressure than fluids. As their viscosity is low, they have great diffusivity and have a better connection with the substrate. The utilization of supercritical CO_2 in material coloring is an ecologically friendly, manageable option. Rather than utilizing a fluid, supercritical CO_2 is utilized. It is nonhazardous, nonflammable, and easily accessible and does not leave deposits. Being a nonpolar particle, it carries on like a nonpolar natural solvent in its supercritical state. Supercritical carbon dioxide (SC CO_2) has been utilized for extraction of high-value mixes from common substances, particularly for food applications as, unlike natural solvents, there is no leftover solvent in the extracted material. In spite of the fact that the idea of SC CO_2 coloring was proposed in the 1980s, its use for viable coloring applications began just in the decade of the 2010s.

The procedure continues in an indistinguishable way from the regular strategy, yet as opposed to sending the spent blend to the waste treatment system, the supercritical CO_2/color blend is depressurized. The CO_2 changes to a gas and all the spent color drops out and can be reused. In production systems, the CO_2 is reused for accommodating a totally enclosed framework and a completely ecologically inviting way to deal with material coloring. The major advantages of using supercritical fluids in the coloring process are that water use is reduced during the dyeing process, water pollution is avoided, the time requirement is low, and a large quantity of water can be saved. Fig. 4.3 presents a schematic of textile dyeing with supercritical CO_2.

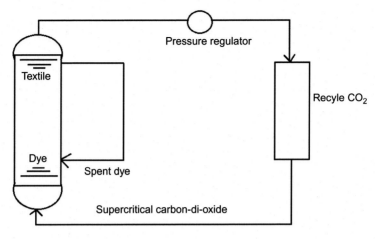

Fig. 4.3 Dyeing process using supercritical CO_2.

Dutch organization, DyeCoo Textile frameworks, was the first to offer businesses a SC CO_2 coloring framework. It collaborated with Huntsman Textile Effects to together create an innovation in supercritical CO_2 material preparation. Adidas was the primary brand to present a product offering utilizing this innovation with the assembly of 50,000 dry dye T-shirts in 2012. This innovation has additionally been found to be helpful in wax expulsion/scouring of cotton materials. Despite the management appeal, the high hardware cost has limited the across-the-board reception of supercritical CO_2 innovation by the material business. Distinctive methodologies, for example, fiber adjustment, color change, and the inclusion of a solvent or a modifier to enhance the solvency of polar solutes, have been attempted to take care of this issue. However, they have not made much progress; hence, more advancements are required for this innovation to be better received.

A lot of wastewater is created particularly in the coloring and bleaching forms, bringing about wastewater that frequently winds up in water bodies. The government intends to expand the growth of textile exports to 15%–20% by 2019. This objective will be accomplished through various key activities, for example, enhancing intensity by redesigning the framework to build work profitability. An investigation that was carried out by the Center for Science and Environment (CSE) highlighted the water scarcity and pollution issues that are caused by the wet processing techniques followed in the textile industry. The investigation assessed that the water utilization by the Indian textile and fashion industry alone is around 200–250 m^3/ton cotton fabric of water in contrast with the worldwide best of under 100 m^3/ton cotton cloth. The amount of water required in material handling shifts broadly indeed, even between comparable kinds of wet handling at various locales. Heating of dye baths constitutes the major part of the energy expended in coloring; hence, utilizing low-liquor proportion coloring hardware can bring about significant energy savings. The washing proficiency relies upon mechanical factors, for example, dyeing baths and texture turnover rate or "contacts," the liquor proportion, turbulence, and other mechanical actions and physical stream qualities. Therefore, bringing down the liquor proportion and reducing aggregate water use are generally not closely associated.

4.3 Global laws for water withdrawal

Apart from creating changes in the manufacturing methods of the textile industry, laws and regulations also play a vital role in the reduction of water withdrawal and water consumption (Pan et al., 2012). This issue highly pertinent for both small and medium organizations and substantial measured manufacturing operations because this hole of learning lead to missing the beginning stages for water minimization attempts (Redmond et al., 2008). Data in regards to water withdrawal and utilization are generally just checked on a manufacturing plant level and assigned the individual process chains by means of assignment rules. Global studies, for example, of the OECD (Organization for Economic Cooperation and Development) additionally forecast an expanded worldwide water request of 55% by 2050, incorporating an expansion in manufacturing by 400%, which is affected by the leading developing countries. Water, which is a valuable asset, is required and there is a need to implement some of the regulations based on water withdrawal and consumption. Textile manufacturers need to embrace water protection and contamination control technologies on account of progressive water shortages and water contamination, particularly in the textile industry (Chen et al., 2017). A powerful execution tool for these technologies is very much needed for textile manufacturers. But the available tools in the textile industry can only make assessments on the plant level and overlook the contamination impact of water. In order to solve the water issues in the world, some of the countries have initiated laws. There is a requirement for decentralization of water regulatory governing bodies. Neighborhood institutions frequently have better information of water issues and, along these lines, ought to be permitted to arrange and mutually choose water-administration methodologies for taking successful measures. Moreover, they must attempt to check water sources and know about remedial measures. Because water is a constrained asset, its sharing and circulation require an authoritative framework. This can be refined by shaping water laws. Laws are expected: (Alkaya and Demirer, 2015) to guarantee that the dissemination of good-quality water is even-handed, (Bach et al., 1998) to shield the water assets from misuse, and (Balachandran and Rudramoorthy, 2008) to guarantee supportability. Moreover, the execution of these laws ought to be checked to substantiate the achievement of them. Recently, a new enactment called the National Green Tribunal Act 2010 was passed for the successful and speedy disposition of cases relating to environmental security. This law is a basic departure from the present authorization in which, separately from closing down a contaminating industry, controlling supply, and criminal discipline for those in charge of running it, enactments are made accessible to the victims of water contamination. In this way, the requirements are moving in stringent directions. In China, the government has assigned Cleaner Production Promotion Law in 2003 and several cleaner technologies to the textile manufacturers. The

European Commission also provided reference documents based on the conventional methods followed in the textile industry. Some researchers have also suggested replacing the conventional techniques with new techniques in order to enhance the reduction of water consumption in the textile and fashion industries. These new techniques include replacement through the use of natural dyes instead of synthetic dyes, by using physical methods instead of chemical methods, and low-water-use methods (Ebrahimi and Gashti, 2015; Zhang et al., 2014, 2016; Gashti and Adibzadeh, 2014; Ebrahimi et al., 2011).

4.4 Standards for water withdrawal and consumption

Water consumption in a water supply regularly alludes to the measure of water or vitality conveyed to an individual property by a service organization. There is a high use of water in the main stages of textile manufacturing. Notwithstanding this, the amount of energy or water provided to a particular zone, could also be considered as the request or utilization for the properties inside that area. This is especially applicable for energy, in which a system meter that measures supply to a locale could also speak to the utilization (or request) for that suburb. From natural water bodies, the water is withdrawn and used (Simons et al., 2016). The water involved in the stages of textile manufacturing methods are sizing, desizing, scouring, bleaching, mechanical finishing, dyeing, and finishing. Different research exercises have been done widely as a result of natural issues and the increase in the cost of water in ordinary coloring forms (Bach et al., 1998; De Giorgi Cadoni et al., 2000; Gebert et al., 1994; Guzel and Akgerman, 1999; Poulakis et al., 1991; West et al., 1998; Youssef, 2000). A flowchart of the textile manufacturing process is shown in Fig. 4.4. The most water-consuming industries in the world are textile industries. For instance, the textile coloring process consumes 100 L/kg. The shortage of water assets and water contamination turn out to be progressively more serious because of the mishandling of water and lack of treatment in time, while the worldwide industrial water necessity keeps up its development in the meantime under financial developments. The inconsistency between these two patterns encourages modern producers to embrace cleaner production advancements to spare water utilization and decrease water contamination.

The greater part of the textile manufacturing process, such as the processes of scouring, washing, dyeing, bleaching, sizing, and finishing, consumes huge volumes of new water and releases expansive volumes of effluent, which are for the most part contaminated with extraordinary color, a higher concentration of organic compounds, and vast varieties in compounds (Takahashi and Kumagai, 2006). Typically, excess water use, considering waste alone in the majority of the units, works out to be 10%–20% on average (Balachandran and Rudramoorthy, 2008). Water use was considered by authorities to find a way to secure the water for the future generations (Jorgensen et al., 2009). Other global associations have built up different natural norms and measures to avoid over withdrawal and consumption from textile manufacturing. The EKO sustainable textile standard, Organic textile standard

Fig. 4.4 Steps involved in the textile product manufacturing process.

(Italy), Organic Fiber Standards (USA), Standards for Processing of Organic Textile Products (Argentina) are some of the widely established standards. The benchmarks set by these associations guarantee reduction of material contamination and advance the use of natural strands and eco-accommodating material generation techniques. Various ecological benchmarks have been established to lessen the natural effects of the material industry. Some broad and essential ones are utilization of common and azo-free colors, utilization of natural strands, limiting carbon impression, a decrease in the use of water and energy, and so forth. The material business discharges a huge amount of unstable chemicals and contamination for the most part from covering, completing the process of coloring, and printing forms. The standards for water consumption by the industry should be developed by considering the water withdrawal from various regions and from scientific research.

4.5 Challenges and opportunities

This chapter aims to help determine the opportunities and limitations for water withdrawal and conservation in the textile and fashion industries. Additionally, it provides the analytical framework to study the impact of water withdrawal and water conservation in the textile sectors. There is great variety in the worldwide textile and fashion industries and this can bring about complex water-related dangers that must be overseen by textile processing operations. The end goal is to moderate or deal with these dangers. Industry requires a larger number of assets than exclusively material assets and energy, which are likewise characteristically connected to other assets (Rio Carrillo and Frei, 2009). One imperative asset that has regularly been ignored in this unique circumstance is water. Water is a basic asset for manufacturing purposes but then the hidden possibilities in upgrading water frameworks and the corresponding tasks have been neglected. Stringent regulations and technology development are crucial to controlling water withdrawal and water consumption in the textile industry, while technology development is highly recommended in the near-term future rather than strict regulations. An attainable innovation in the textile and fashion industries is water reconditioning and reuse to consolidate better water administration and supportability in the textile manufacturing process. Successful use of conservation techniques is based on information concerning the associated risks. With the administration focusing on its "Make in India" objective, the offer of water use by manufacturing divisions will increase. This will then make the undertaking of water preservation altogether more imperative. New production techniques that use less water must be produced and received on a huge scale. By creating appropriate manufacturing techniques and receiving proper measures, the reliance of handling enterprises on water can be reduced. With a change in water productivity, a variety of advantages can be achieved by enterprises, including a reduced carbon impression and powerful ecological authority.

Various opportunities are available to reduce the consumption of water in textile industries. With ceaseless upgrades in synthetic execution, procedures ought to be looked into frequently to guarantee that each stage stays important. Numerous organizations have drastically decreased wash water by diminishing the number of process steps included. Washing and flushing are both vital for reducing contamination levels in the texture to predetermined levels. Because water and waste transfer costs have been low, there has been an inclination to abuse water. Because the cost to utilize water is presently increasing, the advancement of water utilization could pay profits. One conceivable choice is to reduce the use of flush water for lighter shades. Numerous cooling water frameworks are worked on a once-through premise. The subsequent high-temperature water is by and large uncontaminated and can be reused in the process as cosmetics or wash water. Washing/flushing is the most widely recognized task in material handling. Numerous procedures include washing and flushing stages, and in this manner, streamlining of these procedures can save a huge amount of water. Washing and flushing require around 70%–75% more water than alternate stages, for example, bleaching and coloring. Countercurrent washing is a built-up procedure normal on consistent extents. This arrangement of the task can essentially decrease water use. Some steps to reduce the water consumption in batch and continuous operations are:

Winch Dyeing: Dropping the color shower and keeping away from flood washing could decrease water use by 25%.

Beam Dyeing: About 60% of water forestalling floods during soaking and rinsing may reduce utilization. Programmed controls have had high use in recent years.

Jig Dyeing: An extensive variety of reductions, from 15% to 79%, is possible by changing from the practice of overflow to stepwise flushing.

Ceaseless Operation: A 20%–30% savings was achieved by introducing programmed water stops. Counter-current washing proved to be the best technique. Even the washing hardware had equivalent performance as two vertical washing machines, utilizing a similar measure of water.

4.6 Conclusion

The significant target of this chapter was to explore measures and methods that can considerably decrease water withdrawal and consumption in the textile and fashion industries, which rely on a constant supply of amazing water assets. Water problems are of great importance in the world. As discussed earlier, a significant number of arrangements are accessible for different material wet handling activities, which can fundamentally lessen the unfavorable effect on the earth. Be that as it may, there are as yet numerous difficulties and numerous promising advances, for example, plasma and supercritical CO_2 handling need advanced improvements and refinements to make them practical options. It is the obligation of the concerned business and specialists to create suitable choices with the least effect on the environment. The reception of the available procedures for feasible material wet handling has been poor as the vast majority of the manufacturing sites are situated in undeveloped nations and decentralized and these units do not have the assets needed for introduction. Nowadays, water scarcity problems are increasing in many areas in India. Water shortages have two components, withdrawal (intake of water) and consumption (water made not useful for other things). An enormous number of fashion and textile factories are on the agenda while water scarcity remains as a serious concern. Textile manufacturers have been forced to develop some pollution control strategies in order to conserve water and to avoid the high water withdrawal that leads to water shortages and water contamination in the world. Governments subsequently have two parts to play in this situation. In the first place, they have to put down stricter directions and guarantee their appropriate use, keeping in mind the end goal to advance sustainable material generation and to guarantee clean conditions and jobs for the general population. Besides, investors ought to give money-related help to help in modernizing the area also, lessening its ecological impacts. Accreditation plans have contributed toward accomplishing these targets in a limited way. To provide support to the textile and fashion industries, the government has developed few regulations and standards that help them to incorporate or change methods based on product development. For example, textile plants have different types of the production processes. Seven processes are mostly involved at each plant and for each process manufacturers withdraw and consume a higher quantity of water and chemicals. Additionally, the consumption of water is also high. This can be reduced by implementing changes in the process of each plant so that the water withdrawal and consumption gets minimized.

Based on incorporating changes in their methods, textile manufacturers are able to avoid the disadvantages and reduce the negative impacts on the environment. With constant changes in compound execution, procedures ought to be looked into consistently to guarantee that each stage stays vital. Numerous organizations have significantly diminished flush water by decreasing the number of process steps included. By recycling the water more times prior to discharge, the water withdrawal can be reduced. But the consumption is difficult to reduce because production methods and industrial energy are involved. Saving water also involves additional advantages, for example, decreased contamination releases, stronger biological communities, and lower energy requests. As water and effluent costs keep on rising, new advancements for treating and reusing water for process use must be investigated. The water consumption in textile industries can be reduced by monitoring practices, such as turning off running taps and hoses when the machine is not active, conducting reviews to find broken and spilling channels, drums, pumps and valves, the use of programmed controllers for control of fill volume and alcohol temperature in cluster machines, adjustment of pretreatment procedures to suit downstream handling, educating workers on the significance of water preservation. The outcomes of this study will bring out solutions in the textile and fashion industries available in several countries. Changing the attitude of manufacturers about water preservation needs to turn into an industry that will bring a positive and quickened pace of progress. Brands, retailers, and buyers, for their part, should be more conscious and be prepared and willing to pay more for items that don't hurt individuals and nature.

References

Alkaya, E., Demirer, G.N., 2015. Water recycling and reuse in soft drink/beverage industry: a casestudy for sustainable industrial water management in Turkey. Res. Con. Rec 104, 172–180.

Bach, E., Cleve, E., Schollmeyer, E., 1998. The dyeing of pololefin fibres in supercritical carbon dioxide, part I: thermo-mechanical properties of polyolefin fibres after treatment in CO_2 under dyeing conditions. J. Text. Inst. 89, 647–656.

Balachandran, S., Rudramoorthy, R., 2008. Efficient water utilization in textile wet processing. IE J. TX 89, 26–29.

Cai, Z., Qui, Y., Zhang, C., Hwang, Y.J., McCord, M., 2003. Effect of atmospheric plasma treatment on desizing of PVA on cotton. Text. Res. J. 73, 670–674.

Chen, L., Wang, L., Wu, X., Ding, X., 2017. A process-level water conservation and pollution control performance evaluation tool of cleaner production technology in textile industry. J. Clean. Prod. 143, 1137–1143.

Dascalu, T., Acosta-Ortiz, S.E., Ortiz-Morales, M., 2000. Removal of indigo color by laser beam-denim interaction. Opt. Lasers Eng. 34, 179–189.

De Giorgi Cadoni, M.R., Maricca, E., Piras, D., 2000. Dyeing polyester fibres with disperse dyes in supercritical CO_2. Dyes Pigm. 45 (1), 75–79.

Ebrahimi, I., Gashti, M.P., 2015. Extraction of juglone from Pterocarya fraxinifolia leaves for dyeing, anti-fungal finishing, and solar UV protection of wool. Color. Technol. 131 (6), 451–457.

Ebrahimi, I., Kiumarsi, A., Gashti, M.P., Rashidian, R., Norouzi, M.H., 2011. Atmospheric-air plasma enhances coating of different lubricating agents on polyester fiber. Eur. Phys. J. Appl. Phys. 56 (1), 10801.

Gashti, M.P., Adibzadeh, H., 2014. Ultrasound for efficient emulsification and uniform coating of an anionic lubricant on cotton. Fibers Polym. 15 (1), 65–70.

Gebert, B., Saus, W., Knittel, D., Buschmann, H.J., 1994. Dyeing natural fibers with disperse dyes in supercritical carbon dioxide. Text. Res. J. 67 (7), 371–374.

Gleeson, T., Alley, W.M., Allen, D.M., Sophocleous, M.A., Zhou, Y., Taniguchi, M., Vandersteen, J., 2012. Towards sustainable groundwater use: setting long-term goals, backcasting, and managing adaptively. Ground Water 50, 19–26.

Guzel, B., Akgerman, A., 1999. Solubility of disperse and mordant dyes in supercritical CO_2. J. Chem. Eng. Data 44, 83–85.

Jimenez, A., Molina, M.F., Deunff, H.L., 2015. Indigenous peoples and industry water users: mapping the conflicts worldwide. Aquatic Proc. 5, 69–80.

Jorgensen, B., Graymore, M., O'toole, K., 2009. Household water use behavior: an integrated model. J. Environ. Manage. 91, 227–236.

Kan, C.W., Lam, C.F., Chan, C.K., Ng, S.P., 2014. Using atmospheric pressure plasma treatment for treating grey cotton fabric. Carbohydr. Polym. 102, 167–173.

Kiran-Ciliz, N., 2003. Reduction in resource consumption by process modifications in cotton wet processes. J. Clean. Prod. 11, 481–486.

Lee, U., Han, J., Elgowainy, A., Wang, M., 2018. Regional water consumption for hydro and thermal electricity generation in the United States. Appl. Energy 210, 661–672.

Lin, B., Moubarak, M., 2013. Decomposition analysis: change of carbon dioxide emissions in the Chinese textile industry. Renew. Sustain. Energy Rev. 26, 389–396.

Mehmood, T., Kaynak, A., Dai, X.J., Kouzani, A., Magniez, K., de Celis, D.R., Hurren, C.J., du Plessis, J., 2014. Study of oxygen plasma pre-treatment of polyester fabric for improved polypyrrole adhesion. Mater. Chem. Phys. 143, 668–675.

Ondogan, Z., Pamuk, O., Ondogan, E.N., Ozguney, A., 2005. Improving the appearance of all textile products from clothing to home textile using the laser technology. Opt. Laser Technol. 37, 631–637.

Pan, L., Liu, P., Ma, L., Li, Z., 2012. A supply chain based assessment of water issues in the coal industry in China. Energy Policy 48, 93–102.

Poulakis, K., Spee, M., Schnneider, G.M., Knittle, D., Buschmann, H.J., Schollmeyer, E., 1991. Farbung von Polyester in uberkritischem CO_2. Chemiefasern Textillind. 41 (93), 142–147.

Ranganathan, K., Karunagaran, K., Sharma, D.C., 2007. Recycling of wastewaters of textile dyeing industries using advanced treatment technology and cost analysis—case studies. Res. Conserv. Recycl. 50 (3), 306–318.

Redmond, J., Walker, E., Wang, C., 2008. Issues for small businesses with waste management. J. Environ. Manage. 88, 275–285.

Rio Carrillo, A.M., Frei, C., 2009. Water: a key resource in energy production. Energy Policy 37, 4303–4312.

Simons, B.A., Yu, J., Leighton, B., 2016. WESCML: a data standard for exchanging water and energy supply and consumption data. Procedia Eng. 154, 215–222.

Srikanth, R., 2009. Challenges of sustainable water quality management in rural India. Curr. Sci. 97, 317–325.

Sun, D., Stylios, G.K., 2004. Effect of low temperature plasma treatment on the scouring and dyeing of natural fabrics. Text. Res. J. 74, 751–756.

Takahashi, N., Kumagai, T., 2006. Removal of dissolved organic carbon and color from dyeing wastewater by pre-ozonation and subsequent biological treatment. Ozone Sci. Eng. 28, 199–205.

Vajnhandl, S., Valh, J., 2014. The status of water reuse in European textile sector. J. Environ. Manage. 141, 29–35.

West, B.L., Kazarian, S.G., Vincent, M.F., Brantley, N.H., Eckert, C.A., 1998. Supercritical fluid dyeing of PMMA films with azo-dyes. J. Appl. Polym. Sci. 69, 911–919.

World Trade Organization, 2016. World Trade Organization (WTO) International Trade Statistics form 2000–2014. Available from: https://www.wto.org/english/res_e/statis_e/its_e.htm. (Accessed 1 September 2016).

Youssef, Y.A., 2000. Direct dyeing of cotton fabric pre-treated with cationising agents. J. Soc. Dye. Colour. 116, 316–322.

Zhang, B., Wang, L., Luo, L.F., King, M.W., 2014. Natural dye extracted from Chinese gall—the application of color and antibacterial activity to wool fabric. J. Clean. Prod. 80, 204–210.

Zhang, Y.Q., Wei, X.C., Long, J.J., 2016. Ecofriendly synthesis and application of special disperse reactive dyes in waterless coloration of wool with supercritical carbon dioxide. J. Clean. Prod. 133, 746–756.

Further reading

India Infrastructure Report, 2011. Water: Policy and Performance for Sustainable Development. Oxford University Press, New Delhi.

Sharma, D.K., 2015. Water and wastewater quantification in a cotton textile industry. IJISET 2, 288–299.

Verma, A.K., Dash, R.R., Bhunia, P., 2012. A review on chemical coagulation/flocculation technologies for removal of colour from textile wastewaters. J. Environ. Manage. 93, 154–168.

Water footprint management in the fashion supply chain: A review of emerging trends and research challenges

Eirini Aivazidou*,†, Naoum Tsolakis‡
*Laboratory of Statistics and Quantitative Analysis Methods, Division of Industrial Management, Department of Mechanical Engineering, Aristotle University of Thessaloniki, Thessaloniki, Greece, †Department of Computer Science and Engineering, School of Engineering and Architecture, Alma Mater Studiorum – University of Bologna, Bologna, Italy, ‡Centre for International Manufacturing, Institute for Manufacturing (IfM), Department of Engineering, School of Technology, University of Cambridge, Cambridge, United Kingdom

5.1 Introduction

Freshwater constitutes an essential natural resource for the agricultural and industrial sectors, which are responsible for 70% and 22% of the global water use, respectively (UN Water, 2009). In tandem, as freshwater supplies are being depleted at an alarming rate due to rapid population growth, changing lifestyle patterns, and continuous industrialization (Manzardo et al., 2014), projections for 2050 highlight that freshwater requirements for production activities will increase up to 400% (UN Water, 2015) and thus more than 40% of the global population will be living in regions facing severe water scarcity (UN Water, 2014). Notably, over 80% of the worlds' wastewater is released into the environment without any prior treatment (UN Water 2017), further burdening freshwater pollution. With these risks, as environmentally aware consumers become even more sensitive toward water-intensive products (Kang et al., 2017), an increasing number of companies have incorporated water disclosure and stewardship into their corporate responsibility agenda (CDP, 2016).

To evaluate the water-related performance of corporations and products, Hoekstra (2008) introduced the concept of water footprint (WF) as a key performance indicator of the total freshwater volume consumed and polluted directly or indirectly across a product's end-to-end supply chain. Hoekstra et al. (2011) developed a comprehensive methodological framework for quantifying the green, blue, and gray WFs of a product and assessing their impact on sustainability. Specifically, the green WF refers to the rain water consumed by agricultural products, the blue WF to the surface or groundwater incorporated into commodities, and the gray WF to the freshwater required for diluting water contaminated during production. On the contrary, Ridoutt and Pfister

(2010) proposed a life cycle analysis method for WF assessment based on the water use impact in relation to local water stress. From the standardization perspective, ISO 14046 specifies the principles, requirements, and guidelines for WF calculation, evaluation, and reporting to support decision making toward identifying water-related risks and management opportunities (ISO, 2014).

In the fashion industry, sustainability has emerged as a critical issue (Shen et al., 2014) given that textile production contributes to a great extent toward global environmental pollution (Boström and Micheletti, 2016). In this regard, fashion brands, such as Nike, Marks and Spencer, H&M, and Uniqlo, have established environmental sustainability programs (Choi, 2017). Up until the last decade, production activities in the textile industry have been mainly focusing on technological and financial aspects (Ozek, 2017). Nowadays, although cost and lead time constraints, especially in the fast fashion concept, render the adoption of sustainable strategies in apparel supply chains quite challenging (Perry and Towers, 2012), several multinational corporations aim to improve their corporate social responsibility agendas. The latter corporate behavior is in response to the attitude of consumers, who are often willing to pay a premium for an ecofriendly high-quality fashion product (Ellis et al., 2012). For example, the Swedish multinational clothing retailer H&M implements a series of sustainable practices, including the use of ecofriendly raw materials (e.g., organic cotton), the selection of environmentally and socially sustainable suppliers, the reduction of carbon emissions during distribution, as well as the promotion of recycling by engaging consumers (Shen, 2014).

With respect to freshwater appropriation, the clothing industry constitutes one of the most water-consumptive industrial sectors (Franco, 2017), considering that the production of a t-shirt and a pair of jeans may require more than 2700 (WWF, 2013) and 3700 litres (L) of water (Levi Strauss, 2015), respectively. In fact, studies on WF assessment in the textile industry represent 75% and 30% of the relevant research in the light industry and the whole industrial sector, respectively (Aivazidou et al., 2016). Specifically, the cultivation of cotton, which is used in 40% of all clothing produced globally, is the most water-consuming stage across apparel supply chains, further considering that the majority of cotton farming regions face water scarcity (Ravasio, 2012). Except for the agricultural stage, the textile industry utilizes significant volumes of water throughout all processing operations, such as dyeing, bleaching, or washing (Wang et al., 2013). The clothing sector contributes to freshwater pollution due to the chemicals used to process the fabrics (de Brito et al., 2008). From an end-to-end apparel supply chain perspective, the water-related impact of consumers cannot be neglected, particularly in terms of water consumption during laundry and of water pollution due to the use of detergents (Gould, 2014). Actually, laundering uses an average of 1650 L of water per kilogram (kg) of textile (SFA, 2015).

In this work, we aim to advance the research field of sustainable fashion supply chains in terms of freshwater resources' management by providing a framework including all emerging trends, challenges, and opportunities for the production of water-friendly textile products. Overall, the contribution of this work lies in supporting the corporate decision-making process concerning the identification of water-related risks and the development of water-saving strategies across apparel supply chains. The remainder of this chapter is structured as follows. In Section 5.2, all conceptual components and unique characteristics of sustainable fashion supply chains are presented, while an up-to-date review of WF assessment and management in the textile industry is provided. In Section 5.3, we discuss the emerging trends in the clothing industry and their impact on freshwater resources, as well as the research challenges and opportunities for WF management across apparel supply chains. Finally, Section 5.4 closes with the conclusions and future research recommendations.

5.2 Literature background

In this section, we analyze the conceptual background of this research, which is divided into two discrete pillars. In particular, we first provide research evidence on the structure and aspects of sustainable fashion supply chains and we then present case studies of the textile and clothing industries that refer to WF assessment and management.

5.2.1 Sustainable fashion supply chains

From a sector perspective, de Brito et al. (2008) elaborated on the sustainability-driven challenges and contradictions governing the European fashion retail supply chain, while they further discussed the reorganization opportunities of both internal and external firms to manage the emerging sustainability challenges. In addition, Khurana and Ricchetti (2016) reviewed sustainability-driven developments in the fashion business and identified as drivers for the progressive transition of fashion brands to sustainable supply chain management the following: (1) going beyond monitoring, (2) adopting a comprehensive assessment approach, (3) looking beyond the first tier of suppliers, (4) integrating sustainability to core business practices, and (5) bringing transparency to the supply chain. Allwood et al. (2008) articulated a methodology that helps capture scenario analysis of large interventions in an entire sector through integrating the environmental, economic, and social sustainability constituents. The applicability of the proposed methodology was demonstrated through the United Kingdom's clothing and textiles sector, while the study findings pinpointed the critical role of consumer behavior as a main change driver toward truly sustainable clothing and textile supply networks. Drawing on the case of the Indian textile industry, Baskaran et al. (2012) studied and categorized sixty-three suppliers based on six sustainability criteria related to workforce, societal and environmental issues, namely: (1) discrimination, (2) abuse of human rights, (3) child labor, (4) long working hours, (5) unfair competition, and (6) pollution. The study findings indicated that the long working hours criterion is the most prominent in evaluating suppliers in both garment manufacturers and ancillary suppliers, followed by pollution and unfair competition (garment manufacturers) and employing child labor (ancillary suppliers). Recognizing the limitations of mainstream sustainability reporting tools in the fast-fashion industry, Garcia-Torres et al. (2017) proposed a disclosure framework that enables interactive, timely, and dynamic information sharing toward sustainable value creation for the broader supply chain ecosystem. At an operational level, Illge and Preuss (2012) examined the sustainability strategies of H&M, a Swedish multination fashion firm, and Hessnatur, a small niche player in Germany, to adhere to social and environmental standards in the fashion domain. The adoption of a sustainable cotton strategy, that the latter firms embrace, could promote sustainability in the fashion industry owing to three operational constituents: (1) improving conventional cotton production, (2) engaging in organic cotton production, and (3) recycling cotton.

From a supply chain focal point, Macchion et al. (2018) studied the response of Italian fashion companies to increased sustainability requirements and highlighted the necessity to approach the environmental and social sustainability issues from a supply chain perspective. Li et al. (2014) indicated the role of governance to achieve end-to-end sustainability in fast fashion supply chains, while the study findings suggested that a firm's internal and external governance mechanisms should be strengthened to ensure and balance the interest of stakeholders toward the sustainability vision. Amindoust and Saghafinia (2017) proposed a framework for ranking and selecting sustainable suppliers for the textile and clothing industry and the feasibility of the model was exhibited through its application to a case from the textile industry in Malaysia. Furthermore, Caniato et al. (2012) investigated five cases of both small and big firms in the fashion industry to explore drivers, practices, and performance indicators with respect to green supply chain management. The study findings revealed that large companies tend to focus more on product- and process-focused sustainability improvements, while small companies are able to completely reshape their supply chains and apply sustainable inbound and outbound practices. Moreover, Cao et al. (2017) studied sustainability in an end-to-end cotton apparel supply chain in South Africa and confirmed the prevailing role of economic factors over any ecological dimensions. Following a social sustainability point of view, Perry and Towers (2012) identified the inhibitors and the drivers of corporate social responsibility in the fashion supply chain in Sri Lanka. The findings of the study supported that network relationship management could be more essential compared to traditional bureaucratic monitoring and auditing mechanisms. In this context, Norum (2017) supported consumer education to discourage clothing disposition, an essential stage in the overall sustainability impact over the apparel product life cycle. With regard to the life cycle of fashion products, Shen (2014) discussed the operational structure of the H&M supply chain, further stressing the brand's key materials reuse and recycling practices, which result in reduced greenhouse gas emissions along with energy and water savings. In this vein, Chan and Wong (2014) investigated the product-store attributes of eco-fashion in Hong Kong, finding that only store-related attributes positively influence consumers' eco-fashion consumption decisions. Moreover, Sardar et al. (2016) modeled outsourcing contractual arrangements in the global environment, concluding that a combination of domestic and international outsourcing decision making could ensure a sustainable, risk sharing-based strategy. Nagurney and Yu (2012) developed a competitive fashion supply chain network model that captures oligopolistic competition in the case of differentiated products, further considering environmental concerns. The results demonstrated that consumers' environmental consciousness could have a major impact on the profitability level of firms that invest in environmentally friendly technologies for their supply chain operations.

From the standpoint of specific sustainability constraints, Choi (2013a, b) studied the apparel supplier selection problem in the presence of carbon emission taxes, further proving that properly designed taxation schemes could entice retailers to source from the closest, hence, the most environmentally friendly, suppliers in terms of carbon footprint. Considering carbon footprint taxation as well, Hu et al. (2014) modeled a rent-based closed-loop supply chain and benchmarked the sustainability performance of processes, operations, and promotion strategies against the established regular fashion supply chain. Considering the significant environmental impact of the fashion industry, sustainability issues across textile and apparel supply chain networks have received great attention. In particular, Choi et al. (2014) argued that disposable fashion under the fast fashion concept can motivate academic and business-oriented research to explore effective managerial implications and practices around sustainable fashion supply chain management. Fig. 5.1 depicts a first-effort framework of an apparel supply chain in the context of sustainable development, encapsulating the main aspects that should be considered at each supply chain stage to promote environmental and social sustainability.

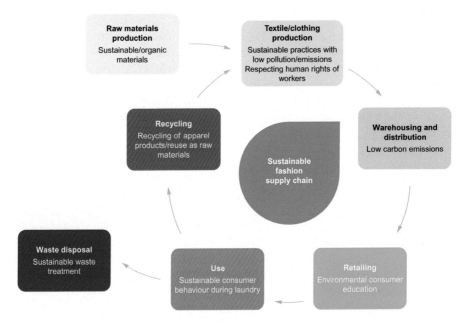

Fig. 5.1 Sustainable fashion supply chain framework.

5.2.2 Water footprint in the textile industry

Water consumption and pollution is not a contemporary issue in the textile industry. Almost half a century ago, Porter et al. (1972) argued that the operations for converting textile fibers into fabrics require a vast volume of freshwater, while they discharge significant amounts of chemicals into water streams. However, the introduction of the WF concept intensified the research effort in the field of water resources' assessment and management across clothing supply chains. To calculate the WF of cotton production in major cotton-producing countries, Chapagain et al. (2006) calculated that the average consumptive WF of cotton is equal to 3644 cubic meters (m^3) per ton (t) of raw material, while that of final textile rises to $9359 m^3$ per t. The authors further highlighted both the crop production and the textile processing impact on freshwater quality due to the fertilizers, pesticides, and chemicals used.

Chico et al. (2013) analyzed the WF of trousers produced in Spain from cotton and wood-based fabrics. The cultivation stage is the major WF hotspot in both cases, while freshwater consumption and pollution during the manufacturing stage differ based on the type of fabric and processing method. In general, wood-based jeans ($1454 m^3$ per unit) are more water friendly than cotton jeans ($3233 m^3$ per unit). Moreover, Joa et al. (2014) developed a novel approach for corporate water accounting, including the water-related performance of different suppliers, which was implemented in a global cotton textile chain. The results revealed that, although the raw materials stage is responsible for the major part of the relevant product WF, manufacturing is also considered to be a critical contributor to the total water use.

Wang et al. (2013) studied the industrial WF of seven knitted fabrics of a textile industry in China, without considering the cultivation stage of cotton. The authors argued that dyeing is the major WF hotspot in the textile industry, followed by bleaching, washing, and drying. In particular, these processes account for more than 95% of the fabrics' WF. Furthermore, Chen et al. (2015) proposed a new tool for evaluating both direct and indirect water use of industrial products. Their methodology was implemented in two typical products of the textile industry, that is, a screen printing fabric and a digital printing fabric. The analysis was based on data provided by a printing and dyeing company in Eastern China. The results indicated that the average water withdrawal per meter (m) of screen printing fabric is 5.4 times that of digital printing fabric.

Rudenko et al. (2013) provided an integrated approach for estimating the financial and freshwater resources of the cotton chain in Uzbekistan through a combination of the concepts of value chain and WF. Actually, the authors considered both microeconomic and macroeconomic analyses of cotton production, processing, and exports in Uzbekistan. The microeconomic results highlighted that the WF of a cotton t-shirt equals $2865 m^3$, while the value added is approximately 0.7 US dollars (USD) per item. From a macroeconomic point of view, the cotton exports amount to 1234 million USD, representing 22% of total export volume, while the related WF rises to $20286 m^3$, constituting 72% of the total WF of exports. More recently, Li et al. (2017) argued that the WF of the textile industry and the economic growth of the sector in China are strongly decoupled, as an increasing annual production value in USD often led to a decreasing annual WF of the sector. These findings may be explained by the water management policies implemented by Chinese clothing industries to control wastewater.

Regarding the production of auxiliary materials for the clothing industry, Zhang et al. (2014) calculated the industrial WF of three types of a typical zipper manufactured in China, that is, one metal and two nonmetal zippers. The outcomes revealed that the metal zipper has the largest WF among the three product types. Painting, dyeing, and primary processing constitute the top three water-consuming processes and contribute about 90% of the total WF. In particular, painting requires the largest amount of freshwater among all processes and is responsible for more than 50% of the zippers' industrial WF. Notably, water pollution is the dominant factor of water use, contributing by 80% to the total WF. Based on the aforementioned review, Fig. 5.2 illustrates the required resources for the main processes of the textile industry, as well as the impact of these processes on freshwater resources, further incorporating the indirect water impact of consumer operations.

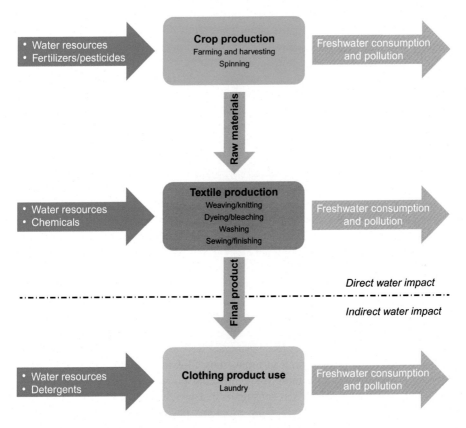

Fig. 5.2 Impact of textile industry on freshwater resources.

5.3 Trends and challenges for water footprint management in fashion supply chains

In this section, we discuss all emerging trends in the fashion industry and the manner in which they affect freshwater resources, from both the academic and the corporate literature. Based on these trends, we then provide research challenges and relevant opportunities for WF management in apparel supply chains.

5.3.1 Emerging trends

Emerging macro-level trends and popularity cues that shape the apparel industry mainly relate to regulatory/institutional, technological, and market developments. In principal, the United Nation's sustainable development agenda identifies seventeen global sustainable goals, among which the access to clean water and sanitation (Sustainability Goal #6) has a prominent role (United Nations, 2016). To this effect, following the consumers' environmental awareness, investors realize that sustainability is essential for ensuring long-term business viability in the fashion industry (Donaldson, 2017). In addition, the circular economy discourse motivates producers, fashion designers, and consumers to consider various aspects, such as sustainable material flow, ecofriendly manufacturing technologies, clothing longevity, and biological and technical cycles to mitigate water use and pollution (Kozlowski et al., 2018). Textiles could have a central role in the circular economy era with an iterative life cycle free of hazardous and/or toxic substances (Nordic, 2015). Nevertheless, the increasingly complex supply networks operated by leading fashion brands dictate the need for the adoption of corporate social responsibility practices, which should be applied catholically across the globally distributed manufacturing sites (Todeschini et al., 2017). In particular, water-focused corporate social responsibility business traits include: (1) ethics in communicating green-washing and water-friendly operations, (2) water-focused sustainability reporting, and (3) supplier disclosure and transparency over water appropriation and efficiency. Corporate sustainability reports confirm that supplier compliance with the fast fashion brands' code of conduct limits the overall supply chain environmental impact (Turker and Altuntas, 2014), further ensuring quality and standards in working conditions throughout the supply chain.

In tandem with the regulatory and institutional advances, consumers' requirements and needs shape the fashion industry landscape and promote sustainability transformations. Indicatively, fashion and apparel is the most popular e-commerce retailing category (Sebald and Jacob, 2018), resulting in a dramatic rise in demand for stylish and affordable clothing. In response, businesses have streamlined their supply chains to reduce costs and production lead times, thus allowing clothing manufacturers to introduce new product lines more frequently in a price competitive manner (Tyler et al., 2006). Moreover, given that consumers treasure environmentally friendly products, brand investments and marketing programs put emphasis on remanufactured fashion operations (Choi, 2017).

Considering the rising clothing consumption levels in emerging economies, the environmental footprint of the apparel industry is expected to expand greatly, hence motivating firms to undertake sustainability-oriented actions (Remy et al., 2016), including: (1) setting sustainable responsibility standards in cotton production and garment design, (2) using nontoxic and sustainable materials, (3) establishing reuse and recycling programs, (4) offering garment repair services, (5) using organic cotton, (6) investing in new fibers, (7) educating consumers about clothing-care practices, and (8) consulting suppliers over labor and environmental issues. Except for the increasing demand patterns, consumers' environmental sensitivity comprises a new driver of the purchasing behavior in the fashion industry (Amed and Berg, 2017). The tendency of consumers to reflect sustainability issues on their persona and relate this notion to the corporate brand image (Allwood et al., 2006) forces fashion firms to invest in sustainable materials and processes, such as the use of organic fabrics, and to strengthen relationships with existing clientele to alleviate environmental consciousness (Amed and Berg, 2017). In this context, campaigns by nongovernmental organizations aim to stimulate consumers' awareness about irresponsibility in terms of sustainability brands in the fashion industry, thus further pressuring firms (Grappi et al., 2017). In this context, the principles of "lowsumerism" (i.e., a commitment to purchase and use a limited number of products) are inducing firms to design fashion products that are timeless and versatile, hence reducing the need for freshwater appropriation in manufacturing (Todeschini et al., 2017). Another emerging trend that could have a positive impact on freshwater resources involves the consolidation of the sharing economy and collaborative consumption (Todeschini et al., 2017). For example, clothing libraries are a typical collaborative consumption business model that allows members to borrow clothes for a specific time period (Zamani et al., 2017). In Sweden, typical cases include the small-scale businesses of Lånegarderoben and Klädoteket. Such sharing, exchanging, and donating practices manifested in fashion supply chains lead to collaborative and sustainable consumption by removing the need for large-scale industrial production (Botsman and Rogers, 2010).

To respond to the emerging trends in the apparel industry, technological innovations and digitalization could promote sustainable solutions across end-to-end fashion supply chain operations (Jhanji, 2018). In terms of materials, organic fibers and alternative synthetic raw materials now offer improved clothing durability, generate reduced waste during cleaning processes, and could be also naturally regenerative (e.g., bamboo), thus limiting the need for consuming virgin resources (Télio, 2015). In terms of manufacturing processes, 3D virtual prototyping allows fashion designers to experiment easily with a variety of fabrics and patterns, thus promoting reduced prototyping lead time, less water and energy consumption, and reduced waste (Papahristou and Bilalis, 2017). Following this, the diffusion of 3D printing manufacturing technologies in the garment industry could eliminate the need for traditional care procedures, such as laundering, which are associated with significant environmental implications (Vanderploeg et al., 2017). In addition, 3D printing offers the potential for more sustainable manufacturing processes, as for example in the subtractive process of fabric cutting, which typically results in significant leftover wasted material (Vanderploeg et al., 2017). Finally, 3D printing manufacturing processes could unveil recycling opportunities, as consumers can be equipped with home-based 3D printing recycling systems to generate printed material from existing 3D printed or plastic-based products, such as polyethylene used in clothes, and use this material in future print jobs (Sun and Zhao, 2017).

5.3.2 Challenges and research opportunities

In terms of water-related sustainability challenges, the textile and clothing industry is governed by four main challenges (Resta et al., 2016), namely: (1) the use of unrecoverable materials and blends, (2) the appropriation and pollution of freshwater resources, (3) the elaboration of hazardous chemical materials and substances, and (4) the miscommunication of corporate social responsibility agendas. To this effect, fashion supply chain reconfiguration opportunities, which are enabled by innovations and aim to promote network circularity, constitute a complex and challenging process (Franco, 2017). The aforementioned challenges should be positively perceived by the stakeholders involved and inform them about interesting grounds for expanding their research and practice horizons.

Firstly, the primary challenge in the textile industry is the use of unsustainable raw materials. In this regard, industry actors experiment with different types of environmentally friendly raw materials (Todeschini et al., 2017), including organic cotton, hemp, bamboo, lyocell, and recycled fibers. In addition, considering that almost three-fifths of all clothing produced is incinerated or landfilled on an annual basis (Amed and Berg, 2017), the use of renewable, biodegradable, and recyclable cotton is another sustainable option. However, the circulation of discarded textiles in principal requires the engagement of producers, operators, and consumers (Dahlbo et al., 2017), while the resulting wastewater reuse in manufacturing operations is challenging, as every fiber has different requirements in terms of reclaimed water quality (Maryam and Büyükgüngör, 2018). In addition, a lag in technological investments and industrial developments hinders the proliferation of sustainable cotton in the apparel industry (Ravasio, 2012). Interestingly, the fibrous matter of agrifood commodities offers a potential opportunity for exploiting alternative fiber sources and developing sustainable textiles, further promoting cooperation between the food industry and the textile sector (Minderhoud, 2015). Indicative byproducts of food crop production and consumption that could provide natural materials for manufacturing fabrics include banana, pineapple, palm oil, and coconut.

Secondly, the cultivation of cotton, which is the most-consumed natural fiber in the textile and clothing industry, requires vast amounts of water (Esteve-Turrillas and de la Guardia, 2017), implying the need for implementing water-efficient irrigation management practices (Feike et al., 2017). More specifically, the cultivation technologies/processes used on farmlands (Feng et al., 2017), along with the adoption of end-to-end supply chain water accounting methods (Joa et al., 2014), need to be comprehensively considered to additionally tackle "not directly visible" freshwater losses through, for example, evaporation and seepage (Cherrett et al., 2005). Following the raw material sourcing echelon of operations, water is a scarce natural resource and a fundamental production factor/medium in most of the textile manufacturing processes (Chen et al., 2017), such as sieving, scouring, washing, desizing, printing, rinsing, bleaching, sizing, mercerizing, carbonization, finishing, and dyeing (Dilaver et al., 2018), owing to its low price and availability. However, the discharged toxic effluents can neither be sufficiently treated nor are they biodegradable (Senthil Kumar and Gunasundari, 2018), hence necessitating novel wastewater treatment technologies and processes that are currently either under development or not yet scalable at an industrial production level. In a similar context, textile engineering research puts emphasis on the development and use of ecological solvents, such as ionic liquid, in key processes that can also lead to more sustainable and innovative textiles with valuable properties (Meksi and Moussa, 2017). However, the adoption of such novel classes of technological fluids in the textile industry remains timid and limited due to their cost and the lack of sufficient in-depth applied research. Furthermore, water-efficient dyeing and washing technologies for natural and manmade fibers, mainly assisted by ultrasound, could also provide feasible options in addressing the increasing water use in the textile industry (Hasanbeigi and Price, 2015).

Furthermore, the use of synthetic, hazardous, chemical dyes and substances by the textile industry in the various processing operations results in the discharge of large amounts of industrial wastewater (Rajoriya et al., 2018). Although several physical, chemical, and biological processes have been developed for the treatment of textile wastewaters, such processes are characterized by the presence of carcinogenic compounds (Slokar and Majcen Le Marechal, 1998). Additionally, such conventional treatment processes produce significant amounts of secondary pollutants, which ultimately increase the environmental impact of a wastewater treatment facility (Lin and Chen, 1997). In this regard, although modern hybrid treatment methods that do not require any chemicals are being explored, their application on an industrial scale has been limited so far due to high maintenance costs (Rajoriya et al., 2017).

Finally, as fashion supply networks involve multiple actors with a high global geographical dispersion, leading brands fail to effectively create awareness about their environmental and social footprint (Eder-Hansen et al., 2017). More specifically, fashion manufacturing is often detached from design, marketing, and consumption, while it is typically relegated to developing countries due to low-cost labor and lenient working and environmental regulations (Todeschini et al., 2017). In this regard, global fashion players should adopt comprehensive corporate social responsibility programs (e.g., voluntary labor standards, sustainability reporting, auditing initiatives, codes of conduct), along with transparency and accountability practices across every supply chain echelon, according to the specific concerns of local communities over local environmental and societal assets (Abreu et al., 2012).

5.4 Discussion and conclusions

Provided that trends in the fashion ecosystem change frequently, intense challenges emerge at the environmental, social, and governance fronts, let alone the economic arena (Retail Forum for Sustainability, 2013). Fashion supply networks are strongly intertwined with consumers' environmental awareness (Caniato et al., 2012) and social consciousness (Huq et al., 2016); hence, as the industry recognizes an accelerating shift of consumer demand patterns toward ecofriendly and socially responsible fashion products, an increasing number of leading apparel retailers have been transforming their supply chains into sustainable ones. This transition toward sustainability includes the implementation of diverse practices, such as the use of organic raw materials, the development of sustainable manufacturing techniques, the environmental education of consumers about prudent laundry activities, and the employment of textile waste treatment, recycling, and remanufacturing techniques. Emphasizing natural resources, the fashion industry constitutes a highly water-consuming production sector (Franco, 2017), taking into consideration the significant direct water requirements of cotton cultivation and textile manufacturing, as well as the indirect water usage and pollution during home laundering and secondary fashion material manufacturing. As such, both the scientific and the corporate communities have been tackling WF assessment and stewardship across clothing supply chains as critical issues for achieving enviro-economic sustainability.

The emerging trends in the fashion supply chain expedite new research opportunities to efficiently study and manage the associated risks and challenges. For example, circular economy, fair trade, "lowsumerism," the sharing economy, and 3D printing are indicative emerging trends in the fashion industry; however, a gap is evident between the associated theory and practice with regard to the resulting environmental and social sustainability impact (Todeschini et al., 2017). From a business perspective, these trends motivate modifications to fashion production patterns and designs (Marsillac and Roh, 2014), while they further promote water stewardship partnerships that aim to implement game-changing strategies in the industry (Li et al., 2014). In addition, as sustainability tends to be a key value and motivation for entrepreneurs, new opportunities for fashion startups are emerging (Todeschini et al., 2017).

The translation of the major trends to valuable sustainability propositions in the fashion industry is associated with key challenges. From a research viewpoint, water appropriation and flows in the fashion industry should be regarded from a life cycle analysis standpoint; however, as fabric materials are sourced from multiple countries, in tandem with the related data constraints, it is challenging to report the associated life cycle impacts (Muthukumarana et al., 2018). From a business perspective, the adoption of water-friendly materials and production processes is still not unanimously perceived as a strategic priority for the fashion industry (Todeschini et al., 2017). Furthermore, educating consumers about garment lifespan, water-friendly washing techniques, and alternative end-of-life uses in order to make informed decisions regarding sustainable apparel selections is rather limited due to cost and capacity constraints (Pookulangara and Shephard, 2013).

To conclude, we anticipate that the proposed conceptual framework, including all emerging trends and water management challenges in apparel supply chains, could act as a roadmap for decision making in the textile industry with respect to the development of state-of-the-art strategies for water-friendly fashion products. Regarding future research, we propose the development of mathematical and/or simulation-based models for the ex-ante evaluation of potential water management policies on the enviro-economic performance of textile production. Furthermore, we encourage the development of real-world case studies concerning the evaluation of the water-related efficiency across end-to-end clothing supply chains, further considering the adoption of the circular economy principals into the fashion industry to support water stewardship.

References

Abreu, M.C.S.D., Castro, F.D., Soares, F.D.A., da Silva Filho, J.C.L., 2012. A comparative understanding of corporate social responsibility of textile firms in Brazil and China. J. Clean. Prod. 20 (1), 119–126.

Aivazidou, E., Tsolakis, N., Iakovou, E., Vlachos, D., 2016. The emerging role of water footprint in supply chain management: A critical literature synthesis and a hierarchical decision-making framework. J. Clean. Prod. 137, 1028–1037.

Allwood, J.M., Laursen, S.E., de Rodríguez, C.M., Bocken, N.M.P., 2006. Well Dressed? The Present and Future Sustainability of Clothing and Textiles in the United Kingdom. Institute for Manufacturing, University of Cambridge, Cambridge. Available at: https://www.ifm.eng.cam.ac.uk/uploads/Resources/Other_Reports/UK_textiles.pdf. (Accessed 10 February 2018).

Allwood, J.M., Laursen, S.E., Russell, S.N., Malvido de Rodríguez, C., Bocken, N.M.P., 2008. An approach to scenario analysis of the sustainability of an industrial sector applied to clothing and textiles in the UK. J. Clean. Prod. 16 (12), 1234–1246.

Amed, I., Berg, A., 2017. The State of Fashion 2017. McKinsey & Company. Available at: https://images.businessoffashion.com/site/uploads/2016/11/The_State_of_Fashion_2017.pdf. (Accessed 10 February 2018).

Amindoust, A., Saghafinia, A., 2017. Textile supplier selection in sustainable supply chain using a modular fuzzy inference system model. J. Textile Inst. 108 (7), 1250–1258.

Baskaran, V., Nachiappan, S., Rahman, S., 2012. Indian textile suppliers' sustainability evaluation using the grey approach. Int. J. Prod. Econ. 135 (2), 647–658.

Boström, M., Micheletti, M., 2016. Introducing the sustainability challenge of textiles and clothing. J. Consum. Policy 39, 367–375.

Botsman, R., Rogers, R., 2010. What's Mine is Yours: The Rise of Collaborative Consumption. Harper Collins, New York.

Caniato, F., Caridi, M., Crippa, L., Moretto, A., 2012. Environmental sustainability in fashion supply chains: An exploratory case based research. Int. J. Prod. Econ. 135 (2), 659–670.

Cao, H., Scudder, C., Dickson, M.A., 2017. Sustainability of apparel supply chain in South Africa: Application of the triple top line model. Cloth. Text. Res. J. 35 (2), 81–97.

CDP, 2016. Thirsty Business: Why Water Is Vital to Climate Action—2016 Annual Report of Corporate Water Disclosure. Carbon Disclosure Project, London. Available at: https://www.cdp.net/CDPResults/CDP-Global-Water-Report-2016.pdf. (Accessed 24 January 2018).

Chan, T., Wong, C.W.Y., 2014. The consumption side of sustainable fashion supply chain—understanding fashion consumer eco-fashion consumption decision. J. Fashion Market. Manag. 16 (2), 193–215.

Chapagain, A.K., Hoekstra, A.Y., Savenije, H.H.G., Gautam, R., 2006. The water footprint of cotton consumption: an assessment of the impact of worldwide consumption of cotton products on the water resources in the cotton producing countries. Ecol. Econ. 60, 186–203.

Chen, L., Ding, X., Wu, X., 2015. Water Management Tool of Industrial Products: A case study of screen printing fabric and digital printing fabric. Ecol. Indicat. 58, 86–94.

Chen, L., Wang, L., Wu, X., Ding, X., 2017. A process-level water conservation and pollution control performance evaluation tool of cleaner production technology in textile industry. J. Clean. Prod. 143, 1137–1143.

Cherrett, N., Barrett, J., Clemett, A., Chadwick, M., Chadwick, M.J., 2005. Ecological Footprint and Water Analysis of Cotton, Hemp and Polyester. Stockholm Environment Institute, Stockholm. Available at:https://www.sei-international.org/mediamanager/documents/Publications/SEI-Report-EcologicalFootprintAndWaterAnalysisOfCottonHempAndPolyester-2005.pdf. (Accessed 21 February 2018).

Chico, D., Aldaya, M.M., Garrido, A., 2013. A water footprint assessment of a pair of jeans: the influence of agricultural policies on the sustainability of consumer products. J. Clean. Prod. 57, 238–248.

Choi, T.-M., 2013a. Optimal apparel supplier selection with forecast updates under carbon emission taxation scheme. Comput. Oper. Res. 40 (11), 2646–2655.

Choi, T.-M., 2013b. Carbon footprint tax on fashion supply chain systems. Int. J. Adv. Manuf. Technol. 68 (1–4), 835–847.

Choi, T.-M., 2017. Pricing and branding for remanufactured fashion products. J. Clean. Prod. 165, 1385–1394.

Choi, T.-M., Chiu, C.-H., Govindan, K., Yue, X., 2014. Sustainable fashion supply chain management: the European scenario. Eur. Manag. J. 32 (5), 821–822.

Dahlbo, H., Aalto, K., Eskelinen, H., Salmenperä, H., 2017. Increasing textile circulation – Consequences and requirements. Sust. Prod. Consum. 9, 44–57.

de Brito, M.P., Carbone, V., Blanquart, C.M., 2008. Towards a sustainable fashion retail supply chain in Europe: Organisation and performance. Int. J. Prod. Econ. 114 (2), 534–553.

Dilaver, M., Hocaoğlu, S.M., Soydemir, G., Dursun, M., Keskinler, B., Koyuncu, İ., Ağtaş, M., 2018. Hot wastewater recovery by using ceramic membrane ultrafiltration and its reusability in textile industry. J. Clean. Prod. 171, 220–233.

Donaldson, T., 2017. The top 7 sustainability trends coming out of textile exchange. Sourc. J.. Available at: https://sourcingjournalonline.com/top-7-sustainability-trends-coming-textile-exchange-td/. (Accessed 10 February 2018).

Eder-Hansen, J., Chalmer, C., Tärneberg, S., Tochtermann, T., Seara, J., Boger, S., Theelen, G., Schwarz, S., Kristensen, L., Jäger, K., 2017. Pulse of the Fashion Industry 2017. The Boston Consulting Group and Global Fashion Agenda. Available at: http://globalfashionagenda.com/wp-content/uploads/2017/05/Pulse-of-the-Fashion-Industry_2017.pdf. (Accessed 11 February 2018).

Ellis, J.L., McCracken, V.A., Skuza, N., 2012. Insights into willingness to pay for organic cotton apparel. J. Fashion Market. Manag. 16, 290–305.

Esteve-Turrillas, F.A., de la Guardia, M., 2017. Environmental impact of Recover cotton in textile industry. Res. Conserv. Recyc. 116, 107–115.

Feike, T., Khor, L.Y., Mamitimin, Y., Ha, N., Li, L., Abdusalih, N., Xiao, H., Doluschitz, R., 2017. Determinants of cotton farmers' irrigation water management in arid Northwestern China. Agric Water Manag 187, 1–10.

Feng, L., Dai, J., Tian, L., Zhang, H., Li, W., Dong, H., 2017. Review of the technology for high-yielding and efficient cotton cultivation in the northwest inland cotton-growing region of China. Field Crop Res 208, 18–26.

Franco, M.A., 2017. Circular economy at the micro level: a dynamic view of incumbents' struggles and challenges in the textile industry. J. Clean. Prod. 168, 833–845.

Garcia-Torres, S., Rey-Garcia, M., Albareda-Vivo, L., 2017. Effective disclosure in the fast-fashion industry: From sustainability reporting to action. Sustainability 9, 2256–2282.

Gould, H., 2014. 10 Things You Need to Know About Water Impacts of the Fashion Industry. The Guardian, London. Available at:https://www.theguardian.com/sustainable-business/sustainable-fashion-blog/2014/sep/04/10-things-to-know-water-impact-fashion-industry. (Accessed 24 January 2018).

Grappi, S., Romani, S., Barbarossa, C., 2017. Fashion without pollution: how consumers evaluate brands after an NGO campaign aimed at reducing toxic chemicals in the fashion industry. J. Clean. Prod. 149, 1164–1173.

Hasanbeigi, A., Price, L., 2015. A technical review of emerging technologies for energy and water efficiency and pollution reduction in the textile industry. J. Clean. Prod. 95, 30–44.

Hoekstra, A.Y., 2008. Water Neutral: Reducing and Offsetting the Impacts of Water Footprints. Value of Water Research Report Series No. 28, UNESCO-IHE, Delft. Available at:http://waterfootprint.org/media/downloads/Report28-WaterNeutral_1.pdf. (Accessed 24 January 2018).

Hoekstra, A.Y., Chapagain, A.K., Aldaya, M.M., Mekonnen, M.M., 2011. The Water Footprint Assessment Manual: Setting the Global Standard. Earthscan, London.

Hu, Z.-H., Li, Q., Chen, X.-J., Wang, Y.-F., 2014. Sustainable rent-based closed-loop supply chain for fashion products. Sustainability 6 (10), 7063–7088.

Huq, F.A., Chowdhury, I.N., Klassen, R.D., 2016. Social management capabilities of multinational buying firms and their emerging market suppliers: an exploratory study of the clothing industry. J. Oper. Manag. 46, 19–37.

Illge, L., Preuss, L., 2012. Strategies for sustainable cotton: Comparing niche with mainstream markets. Corp. Soc. Respon. Environ. Manag. 19, 102–113.

ISO, 2014. ISO 14046:2014—Environmental Management—Water Footprint—Principles, Requirements and Guidelines. International Organization for Standardization, Geneva.

Jhanji, Y., 2018. Computer-aided design—garment designing and patternmaking. In: Nayak, R., Padhye, R. (Eds.), The Textile Institute Book Series—Automation in Garment Manufacturing. Woodhead Publishing, Cambridge, pp. 253–290.

Joa, B., Hottenroth, H., Jungmichel, N., Schmidt, M., 2014. Introduction of a feasible performance indicator for corporate water accounting—a case study on the cotton textile chain. J. Clean. Prod. 82, 143–153.

Kang, J., Grable, K., Hustvedt, G., Ahn, M., 2017. Sustainable water consumption: the perspective of Hispanic consumers. J. Environ. Psychol. 50, 94–103.

Khurana, K., Ricchetti, M., 2016. Two decades of sustainable supply chain management in the fashion business, an appraisal. J. Fashion Market. Manag. 20 (1), 89–104.

Kozlowski, A., Searcy, C., Bardecki, M., 2018. The redesign canvas: fashion design as a tool for sustainability. J. Clean. Prod. 183, 194–207.

Levi Strauss, 2015. Sustainability—Planet: Water. Levi Strauss and Co., San Francisco. Available at:http://www.levistrauss.com/sustainability/planet/water/. (Accessed 5 February 2018).

Li, Y., Zhao, X., Shi, D., Li, X., 2014. Governance of sustainable supply chains in the fast fashion industry. Eur. Manag. J. 32 (5), 823–836.

Li, Y., Lu, L., Tan, Y., Wang, L., Shen, M., 2017. Decoupling water consumption and environmental impact on textile industry by using water footprint method: a case study in china. Water 9, 124–137.

Lin, S.H., Chen, M.L., 1997. Treatment of textile wastewater by chemical methods for reuse. Water Res. 31 (4), 868–876.

Macchion, L., Da Giau, A., Caniato, F., Caridi, M., Danese, P., Rinaldi, R., Vinelli, A., 2018. Strategic approaches to sustainability in fashion supply chain management. Prod. Plan. Contr. 29 (1), 9–28.

Manzardo, A., Ren, J., Piantella, A., Mazzi, A., Fedele, A., Scipioni, A., 2014. Integration of water footprint accounting and costs for optimal chemical pulp supply mix in paper industry. J. Clean. Prod. 72, 167–173.

Marsillac, E., Roh, J.J., 2014. Connecting product design, process and supply chain decisions to strengthen global supply chain capabilities. Int. J. Prod. Econ. 147 (Part B), 317–329.

Maryam, B., Büyükgüngör, H., 2018. Wastewater reclamation and reuse trends in Turkey: opportunities and challenges. J. Water Proc. Eng.. (in press).

Meksi, N., Moussa, A., 2017. A review of progress in the ecological application of ionic liquids in textile processes. J. Clean. Prod. 161, 105–126.

Minderhoud, K., 2015. Towards Sustainability in the Textile Sector? A New Paradigm on Fibre Sourcing. Solidaridad, Utrecht. Available at:https://www.solidaridadnetwork.org/sites/solidaridadnetwork.org/files/publications/SolidaridadResearch-TextileApproach-KatieMinderhoud-2015.pdf. (Accessed 11 February 2018).

Muthukumarana, T.T., Karunathilake, H.P., Punchihewa, H.K.G., Manthilake, M.M.I.D., Hewage, K.N., 2018. Life cycle environmental impacts of the apparel industry in Sri Lanka: analysis of the energy sources. J. Clean. Prod. 172, 1346–1357.

Nagurney, A., Yu, M., 2012. Sustainable fashion supply chain management under oligopolistic competition and brand differentiation. Int. J. Prod. Econ. 135 (2), 532–540.

Nordic, 2015. Well Dressed in a Clean Environment: Nordic Action Plan for Sustainable Fashion and Textiles. Nordisk Ministerråd, Copenhagen. Available at: http://norden.diva-portal.org/smash/record.jsf?pid=diva2%3A819423&dswid=-6701. (Accessed 10 February 2018).

Norum, P.S., 2017. Towards sustainable clothing disposition: Exploring the consumer choice to use trash as a disposal option. Sustainability 9 (7), 1187–1201.

Ozek, H.Z., 2017. Sustainability: increasing impact on textile and apparel industry. J. Text. Eng. Fashion Technol. 2 (5), 76–79.

Papahristou, E., Bilalis, N., 2017. Should the fashion industry confront the sustainability challenge with 3D prototyping technology. Int. J. Sust. Eng. 10 (4–5), 207–214.

Perry, P., Towers, N., 2012. Conceptual framework development. CSR implementation in fashion supply chains. Int. J. Phys. Dist. Logist. Manag. 43 (5/6), 478–500.

Pookulangara, S., Shephard, A., 2013. Slow fashion movement: understanding consumer perceptions—an exploratory study. J. Retail. Consum. Serv. 20 (2), 200–206.

Porter, J.J., Lyons, D.W., Nolan, W.F., 1972. Water uses and wastes in the textile industry. Environ. Sci. Tech. 6 (1), 36–41.

Rajoriya, S., Bargole, S., Saharan, V.K., 2017. Degradation of a cationic dye (Rhodamine 6G) using hydrodynamic cavitation coupled with other oxidative agents: Reaction mechanism and pathway. Ultrason. Sonochem. 34, 183–194.

Rajoriya, S., Bargole, S., George, S., Saharan, V.K., 2018. Treatment of textile dyeing industry effluent using hydrodynamic cavitation in combination with advanced oxidation reagents. J. Hazard. Mater. 344, 1109–1115.

Ravasio, P., 2012. How can we stop water from becoming a fashion victim? The Guardian, London. Available at:https://www.theguardian.com/sustainable-business/water-scarcity-fashion-industry. (Accessed 24 January 2018).

Remy, N., Speelman, E., Swartz, S., 2016. Style that's Sustainable: A New Fast-fashion Formula. McKinsey and Company. Available at: https://www.mckinsey.com/business-functions/sustainability-and-resource-productivity/our-insights/style-thats-sustainable-a-new-fast-fashion-formula. (Accessed 10 February 2018).

Resta, B., Gaiardelli, P., Pinto, R., Dotti, S., 2016. Enhancing environmental management in the textile sector: an organisational-life cycle assessment approach. J. Clean. Prod. 135, 620–632.

Retail Forum for Sustainability, 2013. Sustainability of Textiles. European Commission, Brussels. Available at: http://ec.europa.eu/environment/industry/retail/pdf/issue_paper_textiles.pdf. (Accessed 21 February 2018).

Ridoutt, B.G., Pfister, S., 2010. A revised approach to water footprinting to make transparent the impacts of consumption and production on global freshwater scarcity. Glob. Environ. Chang. 20 (1), 113–120.

Rudenko, I., Bekchanov, M., Djanibekov, U., Lamers, J.P.A., 2013. The added value of a water footprint approach: Micro- and macroeconomic analysis of cotton production, processing and export in water bound Uzbekistan. Global Planet. Change 110, 143–151.

Sardar, S., Lee, Y.H., Memon, M.S., 2016. A sustainable outsourcing strategy regarding cost, capacity flexibility, and risk in a textile supply chain. Sustainability 8, 234–252.

Sebald, A.K., Jacob, F., 2018. Help welcome or not: Understanding consumer shopping motivation in curated fashion retailing. J. Retail. Consum. Serv. 40, 188–203.

Senthil Kumar, P., Gunasundari, E., 2018. Sustainable wet processing—an alternative source for detoxifying supply chain in textiles. In: Muthu, S.S. (Ed.), Detox Fashion, Textile Science and Clothing Technology. Springer Nature, Singapore, pp. 37–60.

SFA, 2015. The State of the Apparel Sector–2015 Special Report—Water. Sustainable Fashion Academy, Stockholm. Available at:https://glasaaward.org/wp-content/uploads/2015/05/GLASA_2015_StateofApparelSector_SpecialReport_Water_150624.pdf. (Accessed 10 February 2018).

Shen, B., 2014. Sustainable fashion supply chain: lessons from H&M. Sustainability 6, 6236–6249.

Shen, B., Zheng, J., Chow, P., Chow, K., 2014. Perception of fashion sustainability in online community. J. Textile Inst. 105, 971–979.

Slokar, Y.M., Majcen Le Marechal, A., 1998. Methods of decoloration of textile wastewaters. Dyes Pigm. 37 (4), 335–356.

Sun, L., Zhao, L., 2017. Envisioning the era of 3D printing: a conceptual model for the fashion industry. Fashion Text. 4 (25), 1–16.

Télio, 2015. Organic and Eco-Friendly Fabrics. Télio Fashion Fabrics, Montreal. Available at: https://www.telio.com/wp-content/uploads/2014/08/Organic-Fabrics-Info-Booklet.com pressed.pdf. (Accessed 21 February 2018).

Todeschini, B.V., Cortimiglia, M.N., Callegaro-de-Menezes, D., Ghezzi, A., 2017. Innovative and sustainable business models in the fashion industry: entrepreneurial drivers, opportunities, and challenges. Bus. Horiz. 60 (6), 759–770.

Turker, D., Altuntas, C., 2014. Sustainable supply chain management in the fast fashion industry: an analysis of corporate reports. Eur. Manag. J. 32 (5), 837–849.

Tyler, D., Heeley, J., Bhamra, T., 2006. Supply chain influences on new product development in fashion clothing. J. Fashion Market. Manag. 10 (3), 316–328.

UN Water, 2009. The United Nations World Water Development Report 3—Water in a Changing World. United Nations Educational Scientific and Cultural Organization, Paris. Available at: http://unesdoc.unesco.org/images/0018/001819/181993e.pdf. (Accessed 24 January 2018).

UN Water, 2014. The United Nations World Water Development Report 2014—Water and Energy. United Nations Educational Scientific and Cultural Organization, Paris. Available at: http://unesdoc.unesco.org/images/0022/002257/225741e.pdf. (Accessed 24 January 2018).

UN Water, 2015. The United Nations World Water Development Report 2015—Water for a Sustainable World. United Nations Educational Scientific and Cultural Organization, Paris. Available at: http://unesdoc.unesco.org/images/0023/002318/231823E.pdf. (Accessed 24 January 2018).

UN Water, 2017. The United Nations World Water Development Report 2017—Wastewater the Untapped Resource. United Nations Educational Scientific and Cultural Organization, Paris. Available at: http://unesdoc.unesco.org/images/0024/002471/247153e.pdf. (Accessed 1 February 2018).

United Nations, 2016. Goal 6 Ensure Availability and Sustainable Management of Water and Sanitation for All. United Nations Water, New York. Available at: https://unstats.un.org/sdgs/files/metadata-compilation/Metadata-Goal-6.pdf. (Accessed 21 February 2018).

Vanderploeg, A., Lee, S.-E., Mamp, M., 2017. The application of 3D printing technology in the fashion industry. Int. J. Fashion Des. Technol. Educ. 10 (2), 170–179.

Wang, L., Ding, X., Wu, X., Yu, J., 2013. Textiles industrial water footprint: methodology and study. J. Sci. Ind. Res. 72, 710–715.

WWF, 2013. The Impact of a Cotton T-Shirt. World Wide Fund for Nature, Gland. Available at: https://www.worldwildlife.org/stories/the-impact-of-a-cotton-t-shirt. (Accessed 1 February 2018).

Zamani, B., Sandin, G., Peters, G.M., 2017. Life cycle assessment of clothing libraries: can collaborative consumption reduce the environmental impact of fast fashion? J. Clean. Prod. 162, 1368–1375.

Zhang, Y., Wu, X.-Y., Wang, L.-L., Ding, X.-M., 2014. The industrial water footprint of zippers. Water Sci. Technol. 70 (6), 1025–1031.

Water footprint in fashion and luxury industry

Alice Brenot*, Cécile Chuffart*, Ivan Coste-Manière[†,‡,§,¶], Manon Deroche*, Eva Godat*, Laura Lemoine*, Mukta Ramchandani[‖,#,**,††], Eleonora Sette*, Caroline Tornaire*

*Luxury & Fashion Management, Skema Business School, Lille, France, [†]Luxury & Fashion Management, SKEMA Business School, Suzhou, China, [‡]Luxury & Fashion Management, SKEMA Business School, Sophia Antipolis, France, [§]Luxury & Fashion Management, SKEMA Business School, Belo Horizonte, Brazil, [¶]Sil 'Innov & Eytelia, Courcelles, Belgium, [‖]United International Business School, Zurich, Switzerland, [#]NEOMA Business School, Reims, France, [**]Founder & CEO Moraltive, Zürich, Switzerland, [††]Professor of Marketing at United International Business School, Zurich, Switzerland

6.1 Introduction

"Sustainable development is a dynamic process which enables people to realize their potential and improve their quality of life in ways which simultaneously protect and enhance the earth's life support systems" (Forum for the future). This quote summarizes the meaning of sustainability. This word was born many years ago and came into common language at the World's first Earth Summit in Rio in 1992. In fact, sustainability means changes in the ways we produce things in order to protect our next generation and in order to preserve our resources. Sustainability appears as a new way of life also, to acquire new and better habits to protect the environment. There are many different points of views about the meaning of this word. There exist many variations and extensions to the basic definition. Today, everyone and any company can and should use sustainability. Sustainability can be used in many sectors, such as ready-to-wear, wine and spirits, jewelry, toys, and real estate, among others[1].

One of the biggest issues for companies today (mostly for ready-to-wear) is water. Water is widely used in an unresponsible manner. Water is wasted and is overconsumed. Thus, to keep the idea of the sustainability, water has to be preserved. We talk about the water footprint. The water footprint is the quantity of water used to produce a good or a service. For example, it requires 15,415 L of water to produce 1 kg of beef. This example shows the impressive quantity of water needed just to produce 1 kg of meat. Furthermore, in France, we eat more than 1.5 million tons of meat per year. The footprint can be evaluated for a company or for a particular country. Countries, even developed ones, don't have the same sensibility to the water footprint,

---

☆ In Memoriam our friends, students and much more, our beloved Cédric Laguerre and Keryan Grimault.
[1] http://www.un.org/waterforlifedecade/water_and_sustainable_development.shtml

for example, in China, the water footprint is not important but it is in the United States. Besides, we can determine three water footprints. First of all, we have the green water footprint, which means the water from precipitation in the root zone of the soil and evaporated, transpired, or incorporated by plants. It affects agriculture and forestry products. Then, we have the blue water footprint, which is water that is sourced from surface or groundwater resources and used in, for example, irrigated agriculture and industrial and domestic water consumption. To finish, the gray water footprint is the total of fresh water needed to assimilate pollutants to meet specific water quality standards (from SAV Miller and WWF dated on 2009)[2].

The water footprint helps us to understand and to evaluate how our limited freshwater resources are being consumed and polluted. We have to understand that water is already rare and we have to be careful with its use. Water is used for many things and for many practices in the world. Moreover, water is not distributed fairly in the world. That is why water is a rare commodity and we have to take care of it. We count 748 million people who still don't have access to a source of water. In July 2010, the General Assembly recognized this disparity and adopted a new resolution that "recognized the right to safe and clean drinking water and sanitation as a human right that is essential for the full enjoyment of life and all human rights." Sanitation is even more challenging with 2.5 billion people of concern. These numbers show large disparities between countries and regions, between rich and poor, and between rural and urban people. We should not forget also that hygiene has to be improved in many countries. Water is a source of hygiene[3].

Focussing on the luxury and the fashion market, it grew in 2017 by 5% to 1.2 trillion euros and sales increased by 6% reaching 489 billion euros in total. Consumers are very attractive in this sector year after year. This market is increasing worldwide driven by strong local consumption and also by strong tourist purchases. Even if buying abroad, the economy also increases. In China, for example, the share of personal luxury goods purchased reached 32% in 2017. The market is different for each country (Japan, China, Europe, United States), but it has increased in all of them year after year. The question is, what about the future? Many studies project that growth will continue at 4% or 5% per year for the next 3 years, with the market for personal luxury goods reaching somewhere between 295 and 305 billion euros by 2020. Online shopping will also increase thanks to new technologies and the new available channels. Consequently, the demand will also grow. In that case, our water footprint is in danger. Issues that we can see today won't disappear overnight. Therefore, we have to find solutions to preserve our water resources and find new ways to produce and to consume[4].

In this chapter, we will discuss various ideas and solutions to yield a better "world" for the next generations. First, we will provide an overview of the water footprint, then we will focus on consumer awareness, and in the last part, we will consider solutions

---

[2] https://www.thegef.org/news/importance-water-sustainability

[3] http://waterfootprint.org/en/water-footprint/what-is-water-footprint/

[4] http://pacinst.org/issues/sustainable-water-management-local-to-global/

6.2 The impact of the fashion industry on water resources

Eighty billion sets of clothes are bought every year. With an increase in the average customer value, the sale of clothes has doubled in the past 20 years. In order to meet increasing consumer expectations, the fast fashion sector is steadily producing more clothes for a cheaper price, while increasing the pressure on suppliers.

The fashion industry is the second most polluted sector and keeps provoking indignation and polemics. Besides, fashion brands, such as Zara, H&M, Forever 21, and Primark, hit the front page after several environmental damages.

In fact, these brands constantly sell new styles according to trends, which quickly become popular and then die out. With 50 small collections per year, the fast fashion industry continually pushes customers to buy new and trendy clothes.

6.2.1 China, the first whistleblower

In 2017, the Chinese government wanted to know the real impact of the textile industry in the country. With the help of the *Institute of Public and Environmental Affairs* and the Chinese NGO, *Natural Resources Defense Council*, a real-time map showing the environmental impact of the fashion industry in China has been developed.

This initiative has spotlighted that the Southeast of China has 70% of its water resources polluted by the textile factories. In fact, by producing 75% of the textiles worldwide, many Chinese factories discharge their chemical products into rivers, emphasizing the water pollution, while threatening the natural resources.

By being one of the most profitable sectors, the fashion sector has a social and environmental responsibility; it has to find a new and greener way to balance creativity, production, and distribution in order to manage its resources in the best way.

6.2.2 Denim

In a recent report, Greenpeace has spotlighted pictures of pollution in the Guangdong province of China, where 80% of the jeans are produced. By collecting a sample of water, five metals have been found: cadmium, chrome, lead, copper, and mercury. In fact, the presence of these metals can be justified by the quite complex process involved while manufacturing the denim. In order to obtain the particular denim texture, the factories have to tan, wash, and launder it with the five chemical metals developed before. Then the factories discharge their wastewater into the Chinese rivers.

But it is not only the fabrication of the jeans that has to be highlighted, personal use and the way people wash them also raise the alarm bell. For example, in 2017, the global population washed jeans after two uses. In 20 years, if the French people use

[5] http://www.globalfootprints.org/sustainability

approximately 619 L of water to wash their jeans, American consumers would use 1049 L of water. 260 million pairs of jeans are produced and sold every year; by washing them every 10 times, the water consumption could be reduced by 75%.

Greenpeace has published a report called: Under Wraps. It has revealed that dangerous substances are discharged from textile factories working for Levi's, Calvin Klein, LVMH, and Gap, into the rivers of Mexico. Thus, the NGO rolled out strong actions in order to alert the Mexican and international authorities. As an example, a 110 m banner "Stop polluting Mexico's rivers" has been put next to a Levi's factory. Through this kind of action, Greenpeace sought the publication of all the chemical components used during the process and the chemical percentage then discharged into the rivers of Mexico.

6.2.3 Cotton

The impact of the fashion industry on water does not concern only the denim process of jeans, it also includes the whole cotton production cycle. Approximately 40% of our clothes are produced with cotton. In order to obtain this soft texture, this tropical plant needs water, warmth, and sun. After rice and wheat, cotton is the third largest consumer of irrigation water in the world. Indeed 2700 L of water is needed in order to launder and tan it. That amount represents the average water consumption of a human adult for 2 years. Thus, from 1960 to 2000, the cotton monoculture in Uzbekistan has dried the whole Aral Sea.

Even though biological cotton has been developed, it represents less than 1% of the global area for cotton production. Although biological cotton uses less water than basic cotton, it needs some chemical treatments in order to obtain the same soft and white texture.

For example, in 2017 the fashion brand Primark had committed to an initiative from environmental experts—the Sustainable Clothing Action Plan (SCAP). The brand was working to collective industry targets, such as reducing the amount of waste in order to diminish the environmental impact of the supply chain and to improve the traceability of the cotton in the supply chain[6].

Thus, the fashion brand helped the farmers looking after their environment by using less water and chemical products while still getting better yields.

Moreover, Primark also worked to make the farming communities stronger by building business skills, particularly among female farmers, by investing in health and other projects. Thus, the brand has launched a sustainable cotton program in the city of Gujurat, in India. The aim was to improve the livelihoods of farmers by increasing their incomes and the results were significant. The cotton quality has improved and the farmers increased their profits by more than 200%.

---

[6] https://www.globalsupplychainlawblog.com/fashion/sustainable-fashion-its-about-more-than-just-being-really-really-ridiculously-good-looking/

6.2.4 The textile industry, an opportunity for the developing countries

With this fast fashion consumption trend, developing countries see the textile industry as a strong contributor to the national economy. It can create new jobs, enrich the population, and support economic growth.

Besides, the biggest changes in the textile industry occurred in the 1960s when new production centers began rapidly springing up in Asia. Many of these centers opened first to service the less capital-intensive clothing industry, and then were used to export earnings from these products to set up their own textile production.

But after this wave of industrialization and public health troubles, such as problems of water intoxication and flood, the developing countries have decided to move to a new, green business model, the slow fashion; steadily combining creativity and global distribution for the brands. In fact, designers, such as Alessandro Michele, the Creative Director of Gucci, spotlighted the problem of pressure to deliver new products at an ever-faster clip. Thus, fashion brands decided to move to more transparency with a green footprint in order to justify the rapidity of the development of the collections and to promote their fashion innovations. For example, AYR, a fashion brand introduced in 2014, is a perfect example of low fashion, specializing in made-to-last basics in luxury materials. The fashion brand promotes small collections of cashmere, crewnecks, and jeans, all with a small ecological and water impact. Its latest find, the Aloe jean, is made from high-quality recycled cotton and washed using just one cup of water, instead of the 2700 L of water needed conventionally[7].

If the fashion industry has made some improvements regarding its consumption of water, it can still improve its economic footprint. As the second most polluting industry, the consumers and the global authorities expect a lot from it. For example, the fashion industry could improve the transparency of its supply chain, the consumption of its energy, and the use of sustainable materials among its collections.

6.3 Water awareness in luxury and fashion

6.3.1 A shift in luxury to more sustainability

The bridge between luxury and sustainability is full of controversial ideas. If luxury and sustainability can sound like two opposite concepts, they also appear more connected than ever before. Indeed, according to Dorothy Mackenzie, chairman of Dragon Rouge, a sustainable branding agency, sustainable luxury is an oxymoron because luxury essentially means "not needed," which goes against thinking about sustainable consumption[8]. Nonetheless, in terms of process and consumption, there is no doubt that luxury can support a more sustainable approach than fast fashion[9].

---

[7] https://fashionunited.com/news/business/6-sustainable-textile-innovations-that-will-change-the-fashion-industry/2017100917734
[8] https://www.ft.com/content/5133c2fa-bf90-11df-b9de-00144feab49a
[9] http://about.hm.com/en/sustainability.html

The increasing consumer awareness of sustainability is pushing the industry to review its water footprint and environmental impact, which is creating some shifts in the paradigm and in the image of luxury brands. Lochard and Murat (2011) have already made the assumption that the luxury industry will become greener because of luxury customers extending their expectations of high quality to include environmental factors. (Cassaniti, 2015–2016).

Additionally, the WWF-UK authors Jem Bendell and Anthony Kleanthous stated that today, many luxury consumers are part of an affluent, global elite that is increasingly well educated and concerned about social and environmental issues (Keinan and Crener, 2015). According to them, these luxury consumers consider luxury as a symbol of success. These "successful luxury consumers" understand the meaning of success as a symbol of their wish for a better world (Keinan and Crener, 2015). Moreover, this is not only true for the western population as this phenomenon appears to be growing elsewhere as well. The head of luxury goods research at Citi, Thomas Chauvet, noticed similarities within Chinese and Japanese consumers who feel the need for more transparency in the supply chain (Keinan and Crener, 2015). The rise in consciousness to more sustainability appears to be a global phenomenon. There are different reasons for a consumer to consume in a sustainable way. Even among the environmentally concerned consumers, the reasons to buy ethically differs depending on the profile. Indeed, environmentally concerned consumers displayed different forms of behavior, from a conspicuous or activist commitment to more discreet and individual practices. These consumers also expressed two main motivations: altruistic (e.g., protection of the environment, contribution to the greater good, rejection of market domination) and individualistic (e.g., protection of their own environment, health, and wellbeing; search for the best quality products) (Keinan and Crener, 2015). In addition to this, research has shown that being an environmentally concerned consumer is a combination of gaining global environmental benefits (e.g., reducing greenhouse gas emissions and conserving water or energy) as well as gaining personal benefits derived from social status (e.g., the distinctive personal advantage of "being green") (Keinan and Crener, 2015).

Obviously, authenticity is a crucial matter for environmentally concerned consumers, and this is particularly important in luxury. The perception of consumers depends on the way the brand communicates. Some elements of communication can make the consumers feel more confident about the sustainability of a luxury brand. A study conducted in France revealed that when a brand was perceived as "showing luxury" (overuse of the logo or mass production), consumers had doubts on the legitimacy of the sustainability claims. By contrast, whenever a brand was perceived as genuinely "making luxury" (tradition, craftsmanship, use of rare materials) consumers were more inclined to associate luxury and sustainability (Keinan and Crener, 2015). This is the concept of the "fallacy of clean luxury," which means that, by their very nature, luxury products are perceived to be sustainable (Davies et al., 2012, p.46).

Luxury brands and groups understood this need for more sustainability, even in communication, and global efforts have followed. As an example, in 2013, the renaming of PPR as Kering was a strategic action showing its CEO's philosophy. The naming was not chosen by chance but to show that Kering actually cares.

François-Henri Pinault understands that transparency is the future, and good business practices will be imperative to survival (Keinan and Crener, 2015).

Consequently, the group positioned itself as a leader in sustainability and has launched actions, such as a sustainability department, ethical charter, and a code of conduct, to show its implication and concern.

With luxury companies and groups taking serious actions toward sustainability, the definition of luxury itself will soon, if not yet, be obsolete. Today, one of the challenges in sustainability is actually to use less water and create more value to the luxury good and service. In other words, that would mean consume less but make it last longer. Following this new philosophy or definition of luxury, Bang & Olufsen, the luxury sound system firm, has already adopted this approach and is willing to develop products for a life cycle of up to 15 years (Cassaniti, 2015–2016).

To go even further in this luxury philosophy, Patek Philippe, the luxury watchmaker family company has already understood this concept as well, creating watches while thinking of future generations. The heart of this luxury philosophy appears in their tagline: "You never truly own a Patek Philippe, you merely look after it for the next generation."

If a craft passed from generation to generation is a synonym, somehow, to both a timeless luxury and a sustainable luxury, there are obviously other luxury industries in which transmission cannot even be considered and reviewing the sustainable production process becomes an obligation.

6.3.2 Luxury companies educating consumers

According to experts, predictions show that by 2025 more than half the world's population will be facing water-based vulnerability. Each luxury industry has its main issues. The appropriate solutions for fashion would not be the same for hotel and tourism, for instance. Tourism and hotels are increasing their water consumption and this impact cannot be ignored. Fortunately, companies are taking actions involving consumers. Involving consumers is a means to educate them. Starting with hotels, this is the case of the Radisson Blue with its "Blue Planet" initiative. In 2017, the hotel started a partnership with the international charity "Just a Drop" in order to reuse towels, which has provided more than 5000 children with drinkable water for life. In 2018, the focus is on the "Super Breakfast" which is better in quality and generates 20% less food waste. Consequently, they have reduced both their water and carbon footprint, data that has been confirmed by the Water Footprint Network (WFN) who compared their old and new breakfast. Their new breakfast permitted a reduction of 22% in the water footprint thanks to a considerable reduction in the freshwater used in producing food. To keep preserving water, the Radisson Blue acquaints guests with the topic and even more particularly with the "Blue destinations," such as Cape Town. With its Mediterranean climate and dry summers, water is even more of a challenge so small actions, such as the removal of bag plugs, reduction in laundry, and changing the linen less frequently, are necessary incentives. Communicating on the website of the hotel and distributing informative documents within the hotel during the stay of the guests are ways to communicate internally and to educate consumers.

Another way luxury companies can educate consumers on this topic is by being simply transparent, publishing reports on their website. An example is Kering, which created an EP&L, an Environmental Profit and Loss account to measure their footprints, understand their impact, and translate this into a monetary value. Kering has been for the third time announced as the Industry Leader in the 2017 Dow Jones Sustainability Index (DJSI) in the "Textiles, Apparel & Luxury Goods" sector. Kering is planning via its Strategy 2015 to act on all supply chain environmental impacts with a goal of reducing the Environmental Profit and Loss (EP&L) account by at least 40%, including the remaining carbon emissions and beyond that to also include water use, water and air pollution, waste production, and land use changes (Kering, 2018).

To that end, the following actions will be implemented:

- Enhancing EP&L;
- combining the planetary boundaries concept to further integrate the latest scientific thinking into natural capital accounting;
- integrating more real-time data;
- broadening the scope to include the impact of a product's "use" and "end-of-life" phases.
- expanding offsetting commitments to include a new "insetting" approach to ensure actions across the supply chain deliver climate benefits as well as social value.
- scaling sustainable solutions and synergies.

As a counterexample, the family companies (Armani, Hermès, and Chanel, for example) are not required to publish financial data or strategic information. Consequently, it is to their advantage to seek to maintain their dream image. Knowing that some brands are suffering from greenwashing, the family companies actively seek to maintain the dream. The silence can prevent them from boomerang effects (Cassaniti, 2015–2016).

Finally, a luxury company could use the most common method to educate consumers about luxury and sustainability, namely, advertising and communication. Nonetheless, one needs to beware of greenwashing once again, which can negatively affect the image of the company. By nature, luxury brands communicate less than regular brands. Following the concept of a luxury product, luxury communication isn't widespread. This is also a question considered by Stella McCartney, although sustainability has been a central element of the brand's DNA from the start, this dimension has never been a fundamental aspect of the brand's advertising and communication campaigns. McCartney had wanted to make a name for herself in fashion based on the quality, aesthetics, and desirability of her designs. However, she was beginning to question this stance. On the one hand, she was pondering that a new approach could perhaps be more consistent and attractive to consumers with an increased ecological conscience[10]. On the other hand, she wondered whether it was the right time to alter the company's communication strategy. Many fashion brands were jumping on the sustainability bandwagon with questionable levels of sincerity. What could be the risks and benefits? (Keinan and Crener, 2015)[11].

---

[10] https://www.terraeco.net/La-bonne-conscience-de-H-M-ne,43957.html

[11] http://modissimo.fr/bestof-hm-conscious-2013-greenwashing-ou-reel-progres.html

The fashion industry is the second most polluting industry in the world. It is responsible for 10% of CO_2 emissions. The concept of "fast fashion" especially is one of the worst production concepts.

6.3.3 Awareness of water footprint regarding the consumer of fast fashion

"Fast-fashion is a way to describe inexpensive designs that move quickly from the catwalk to stores to meet new trends." It is possible to have new lines and products many times in a week. By following this kind of trend, consumers buy more and more clothes without thinking. Sometimes it can be a useless or impulsive purchase. This over-consumption is a huge problem for the environment.

Products are fashionable only during a very short time. People throw away their clothes quickly to buy new ones that are trendier. Of course, low prices encourage this mode of fashion consumption. The system increases fashion waste considerably. In France, 70% of disposed clothes end up in a garbage dump. Indeed, to obtain a low production price, firms choose to use poor-quality materials. These products have a limited lifetime, so they are generally not recyclable. The European Union produces about 5.8 million tons of textile waste. Materials from clothes are very polluting, for example, nylon requires between 30 and 40 years to deteriorate. Furthermore, during the production process, textiles are made with many chemical products that are not degradable. Every year, 1.7 million tons of chemical products are used during the dye process. It is polluting the air and groundwater. According to the World Bank, the "fast fashion" industry could be responsible for 17% or 20% of the water contamination around the world. In fact, companies need between 2700 and 7000 L of water to produce one pair of jeans[12].

Thanks to the sustainable trend, people care more and more about the environment. Furthermore, on April 24, 2013, the collapse of Rana Plaza in Bangladesh brought the fashion industry environmental impact to people's attention. The documentary "The true cost" by Andrew Morgan, which exposes the hidden side of the textile industry, strengthens consumer awareness. Every year since the accident, the Fashion Revolution Day rallies consumers thanks to social networks in order to learn more about the textile fabrication process. Last year, the #fashionrevolution was ranked first in the world by Twitter. To keep selling, fashion industries must follow the sustainable trend by showing an ecofriendly brand image. How do they promote their eco-responsible approach? Do the companies really care about the environment?

6.3.4 Fast-fashion industries strategies

H&M, Zara, and Topshop all decided to adopt an ecofriendly brand image. H&M is one of the pioneers in sustainability. Indeed, the group set up many strategies. It created a program in which customers receive a discount on their next purchase when

---

[12] http://www.parismatch.com/Vivre/Mode/L-ecologie-on-dit-oui-tout-de-suite-Une-mode-ethique-et-belle-810533

they return their old clothes. According to the group, in 2014, about eight tons of clothes were collected. Half of the products are reused by other people. One third are recycled and the rest are transformed to rags or used to create energy[13].

The brand also chose to have a transparent policy; in fact, it is one of the first companies that has adopted Human Right Principles from the United Nations. It has also made public its list of suppliers and given information about them. In April 2017, the brand received the Freedom House Reward for its supply chain transparency. H&M launched a line made with used clothes, which were collected in their stores. The products are made with 20% recycled textile. It is not much, but the group announced that it would like to increase these lines by 300% in a year. The process is very complex, and H&M decided to work with PUMA (Kering group) and the startup Worn to find an easy way to recycle clothes into textile products. In every year since 2011, the brand has launched a capsule collection named "Conscious." It is made with a 100% recycled fabric. The company uses nylon accessories, such as fishing nets[14].

More specifically on the topic of water, H&M wants to take responsibility through the whole value chain. Indeed, it has decided to focus on five areas: increase water awareness, communicate more about the impact on water from the value chain, improve water management, sensitize external stakeholders, and influence government. The company chose to fix some objectives in order to improve water conditions in Bangladesh, but also China[15].

With the aim of increasing water awareness, H&M trained its employees with a sustainability e-learning. Workers who buy and design also received information about raw material choices in order to make sustainable purchases in the future. The firm wants also to sensitize its suppliers. In fact, suppliers who use humid processes, such as dye or washing, must treat wastewater by following Business for Social Responsibility Water Group criteria. This is part of the suppliers' audit program. Furthermore, the group made a forbidden dangerous chemical products list that suppliers must respect. It is regularly updated. Since 2013, the brand has also proposed a positive list of good products in order to help their supplier with chemical product choices. Since 2013, H&M has collaborated with Bangladesh local governments and surrounding communities in order to improve wet processes in the textile industry. Thanks to the same partnerships, the group has improved the water management about river basins. This is also the case in China, where H&M works with the World Wildlife Fund (the leading organization in wildlife conservation and endangered species) on a basin project to improve water quality and accessibility.

To remain competitive, the other fast-fashion companies must follow the sustainable trend. In November 2015, Zara launched the "Join Life" collection. Products from this line have many qualifications. Firstly, components must be ecofriendly: the main fabric is organic cotton or cotton approved by the Better Cotton Initiative.

---

[13] https://www.courrierinternational.com/article/enquete-hm-zara-topshop-la-fast-fashion-un-fleau-ecologique

[14] http://www.leparisien.fr/laparisienne/mode/conscious-collection-la-mode-ecolo-d-h-m-03-02-2011-1299913.php

[15] https://madeinresponsable.com/

The brand also uses tencel, recycled polyester, and polyamide or lenzing modal (fibers extracted from wood). Fabrics must be made with "Green to Wear" technologies, including recycled water and with minimal utilization of chemical products by industries. Like H&M, the group decided to practice transparency by making public information about its wet processing units around the world. Zara is going further. In fact, it decided to have "eco-efficient" stores. The main goal of these stores is to decrease electricity consumption and to use greenhouse gas emissions. Compared to traditional stores, these ones consume 30% less energy and 50% less water. In 2015, the brand decided to work on sustainability with its suppliers; the objective is zero landfill waste in 2020. They have already started to recycle and reuse their products (thanks to a "box in stores system" as H&M) but still work with partners, such as universities and others fashion companies, to find an unrecycled product solution. Regarding recycling, the firm makes its packaging into recycled paper and promotes this with a "Boxes with a Past" campaign. Thanks to these plans, the Zara Company would like to fulfill some objectives for 2020. The brand would like to eliminate undesired chemical products, reduce energy consumption by 15% during the manufacturing process for each clothes product on the market, decrease the energy use of its stores by 10% for each clothes product on the market and have 100% of their stores be ecofriendly.

Zara has a water strategy to reduce its water footprint. Based on H&M policy, the group wants to manage and promote water consumption at all supply chain levels. Plans must impact all headquarter offices, logistics centers, and stores.

The British brand Topshop chose the same strategy with the launch of its eco-fashion capsule collection: Reclaim. The company decide to work with Reclaim to Wear, a brand that creates fashion products with industry's surplus stock, such as remnants and off-cuts. Thanks to this partnership and the sharing of knowledge, Topshop made a collection with discarded materials, such as jersey, cotton, and denim that already exist. The Arcadia group cares about the environment, but also about worker wellbeing. The company has a code of conduct that manufacturers have to follow. It is based on the International Labor Organization conventions and recommendations, which ensure that employees are not exploited and promotes decent working conditions. Like its competitors, the brand is a member of the Better Cotton Initiative. The company seeks to have a sustainable image, specifically in the United Kingdom. Indeed, the brand is recycling 95% of its waste into British stores; it is also reducing water and energy use in warehouses, offices, and stores. Finally, in the United Kingdom, the firm uses 100% renewable energy[16].

6.3.5 Green advertising campaign: Reality or greenwashing?

To maximize success, companies must support their sustainable strategies with a green communication campaign. Generally, it is similar for all fast fashion brands and we here choose H&M as an example. For its "Conscious" line, H&M launched a print and TV advertising campaign based on nature and green. The firm used famous

---

[16] http://www.topshop.com/en/tsuk/category/responsibilities-22/home

people, such as Vanessa Paradis, to catch the attention of consumers. The choice was well considered; in fact, Vanessa's brand already had a bohemian and ecological image. The ad shows a fresh atmosphere and blue and green clothes are shown. Of course, vitrines, websites, and social networks convey the same image. Some stores even promote the eco-friendly collection by exchanging the original logo for a green one. Unfortunately, all of these promotions and sustainable strategies could be only a way to improve the brand image of fast-fashion companies and their benefice. To continue with the example of H&M, the leader of this movement, the brand is not lying about the organic components of its products, but there is an offset between the small quantity of "Conscious" clothes and the huge communication campaign. The space in stores for this organic collection is restricted compared to the store area. Moreover, organic and better cotton represent only 11.4% of the whole cotton used by the company. Also, some objectives have only a deadline but not much information and methods are communicated about it. On its progress report, the company shows that many points must still be improved. By communicating a lot about these points, we can say that fast-fashion companies are using greenwashing[17].

To conclude, H&M use a small part of its production to create an eco-friendly brand image. The company supports its image thanks to a huge advertising campaign, which is disproportionate and not even always real. In fact, in 2010, 30% of the organic cotton used by H&M was in reality GMO cotton. About ethic and employees' wellbeing, the company still uses some production methods, such as blasting units that are very dangerous for human health.

Even if the huge advertising campaigns about eco-friendly collection catch the attention of some customers, media and especially social networks facilitate information transmission. They are used as a weapon to denounce the hypocrisy of some fast fashion industries and warn consumers.

6.4 Solutions

As we previously explained, today the overuse of fabrics, such as cotton, which is a highly water-consuming fabric, or petroleum-based fibers, such as polyester, nylon, acrylic, and spandex, are continuously damaging the environment and are increasing the water footprints in every luxury or fashion industry. Countries that produce such fabrics, such as the United States, India, and China, are today facing the challenges of high pollution.

If fashion, and moreover luxury, are based on forecasting future trends, it might be important today to consider new ways of producing thanks to new technologies, science improvements, and maybe effectuate a return to "more natural" fibers, as consumer environmental awareness is growing.

In this part, we will introduce potential solutions to reduce the water footprint challenge that the whole fashion industry is facing.

First, change must occur in the use of raw materials.

---

[17] https://ethique-sur-etiquette.org/H-M-UNconscious-Collection,280

6.4.1 A return to the organic

Cotton represents today a quarter of all the pesticides used in the United States and 25% of all the pesticides used worldwide. But it's also the type of crop that consumes the most in terms of water. Even though the production process of cotton is made in order to keep the production high and keep the prices low, it has become a real issue.

That is why more and more brands are trying today to develop "organic" cotton. In 2004, the Organic Trade Association survey showed that the sales of organic cotton increased by almost 23% over the previous year. Since that time, it has continued to grow. Even big retailers, such as Walmart (the largest American retailer), have started to sell products made with organic cotton, making them at the same time also the world's largest buyer of organic cotton.

Organic cotton means no chemicals used for the last 3 years in a cultivated field, and no intense exploitation. Today, organic cotton helps poor areas, such as Ceara in Brazil, to develop sustainable activities where before everything was about intensive production as the only way to earn a living, even though the fields were being put in danger in the long term. Nevertheless, there has been some reluctance to use this fabric, mostly because of its price and its sensitiveness. Indeed, 1 kg of organic cotton can cost 90% more than the average price of regular cotton on the market, and because it's not grown with chemicals, fields and harvests are sensitive to variations in the weather.

International brands, such as Veja shoes or even Stella McCartney, have made the bet to use such a fabric.

6.4.2 Alternative fibers

It's another problem when it comes to manmade raw materials, such as polyester, nylon, acrylic, and even viscose, which are chemical fibers. Viscose, for example, can be used as a luxurious feeling fabric. It is produced from intensive irrigation monocultures of bamboo or eucalyptus and lots of chemicals, such as carbon disulfide or chlorine, are used in the production process. That's where the EcoVero alternative fabric steps in. It's today the most responsible response available concerning viscose. It's made from sustainable wood (certified in Europe by controlled sources, such as the FSC, or Forest Stewardship Council). Then the cellulose fiber is produced thanks to a process that creates 50% less emissions and use half as much energy and water. The pulp bleaching is even chlorine free. Moreover, the raw material collection and the transformation process are realized on site, thus reducing emissions significantly[18].

Concerning polyester, a report from the Water Footprint Network and C&A (Freitas et al., 2017) shows that the blue water footprint for polyester primarily occurs during the fiber manufacturing stages, while the gray water footprint comes from all production phases with the oil exploration and refinery phases contributing the largest share. That is why some brands use recycled polyester from PET bottles.

[18] https://www.nytimes.com/2017/11/12/style/alternative-fabrics-sustainability-recycling.html

Flax fabric is also perhaps the most promising alternative to cotton. Usually a flax plant is used to make linen. It is pest resistant, it doesn't need a lot of fertilization or chemicals, it needs little to no water irrigation, everything can be used in the plant, and it is recognized as one of the strongest and most durable fibers in the world.

6.4.3 Leather improvements

Leather, being one of the most used materials worldwide, has been scrutinized for its high pollution and environmental impact. It is not only that animals have to be slaughtered for leather sourcing, but also the tanning process with hazardous chemicals is an important issue because it not only has consequences for humans and the environment, but it is also carcinogenic, persistent, and uses an indestructible chemical[19]. It's hard to consider that leather will ever be organic. To supply all the fashion industry, the leather industry is made of huge exploitations with big investments and it's difficult to have control over what type of animal is used to produce the leather and how the animal is treated. Furthermore, the breeding of animals is an over-consuming water industry. Thus, sustainable solutions can be considered[20].

Concerning the leather itself, there is the alternative in another type of leather, the fish skin. It is a good alternative to exotic leather, such as alligator, crocodile, snake, or lizard, which are endangered species, and it is very resistant and durable. We are talking about fish, such as salmon, galuchat, perch, or carp. Those skins that are used usually come from fish used for alimentation purposes and whose skins would have been wasted anyway. Furthermore, the tanning process for those type of skins uses less chemicals than for standard leather. As an example, it does not use any lye or acid, which are great causes of atmospheric pollution. Such leather is used by brands, such as Dior, mainly for jewelry and all leather goods.

Another solution is to change the way standard leather is tanned. Vegetable-tanned leather is an old artisanal way of producing leather, which is free from hazardous chemicals. The tanning is accomplished using tannins from tree barks, such as oak, chestnut, and others (Wikipedia). The time required for tanning vegetable-tanned leather is much greater, as it takes about several weeks and months to complete, which consequently makes it more expensive than the regular chromium-tanned leather. Although many leather industrialists make the criticism that vegetable tanning utilizes more water than chromium-tanned leather, researchers have found that the wastewater can be reused and has a less hazardous impact on the environment compared with chromium-tanned leather. Additionally, there are major advantages of it being a sustainable material as the tanning is more natural, the labor forces are highly trained and experienced, it has high durability and strength, a unique "patina" is acquired over the years of using it, and it can be easily biodegradable compared to chromium-tanned leather (Fernandes, 2014)[21].

---

[19] https://www.consoglobe.com/
[20] http://www.bain.com/publications/articles/luxury-goods-worldwide-market-study-fall-winter-2017.aspx
[21] https://www.zara.com/fr/fr/durabilite-11449.html

6.4.4 New fibers

In recent years, new types of fabrics have emerged thanks to advancements in new technologies. Here are the most promising ones:

Stella McCartney (pioneer in the sustainable fashion luxury) developed a new type of silk in which worms are not needed anymore: the "Laboratory made" silk. It has almost the same texture as silk, but it has been created thanks to recent discoveries in the science of replication of living organisms. One golden dress made out of laboratory-made silk is part of the exhibition of the Museum of Modern Arts.

Recycled fruits are also a promising new land of experiments. An Italian company named Orange Fiber made it its business[22]. As their website explains, they create "exquisite sustainable textiles from citrus juice by-products that would otherwise be thrown away, representing hundreds of thousands of tons of precious resources." This new fabric is created from a silk-like cellulose yarn and can be mixed with other materials. According to the design needed, the 100% citrus textile (known to be lightweight and soft) can be opaque or iridescent. Stella McCartney and Salvatore Ferragamo have used it for their collections. Thanks to that innovation, Orange Fiber won the Global Challenge Award, an innovation challenge held by the nonprofit H&M Foundation in August 2015. Pineapple is also an interesting idea. Presented as the "vegan" leather, pineapple leaves are used to create a natural leather called "Pinatex" developed by the company "Ananas Anam." Pineapple leaf fibers used in this "decortication" process are a byproduct that comes from the pineapple harvest in the Philippines. Even the byproduct created during the manufacturing process that transforms the pineapple leaves into leather (a process called decortication) is converted into organic fertilizer or biogas used by farmers, making it a true sustainable option. LVMH through their sustainable fashion brand "Edun" create items from such a fabric.

A Taiwanese textile company named "Singtex's technology" developed a process to combine coffee grounds with polymer to create a coffee yarn. This yarn possesses such qualities as UV ray protection, quick drying time, and even antismell effects. The company collects and recycles coffee grounds from famous worldwide coffee brands, such as Starbucks.

In the same range of products, banana fibers are also a great deal. Made from the stem of the banana tree, the fiber is strong, biodegradable, and lasts over time. It's known to be even better than bamboo fiber regarding its spin ability, thickness, and strength.

Other possibilities include lotus fibers, stinging nettle fibers, hemp fibers, algae, and the list goes on. Bio-fabricated materials and recycled items have a real impact on saving water. They could be the future of fashion and luxury. But saving water will only be really effective if the consumer changes consumption patterns. Indeed, any expert agrees to say that a more "reasonable" consumption and a conscious awareness of the alternatives possible are the only things that could impact and make the fashion industry adapt itself to producing more durable and sustainable clothes.

---

[22] http://orangefiber.it/en/

6.5 A favorable context

In 2015, the COP21 was held in France, in which 195 countries set a common goal to reduce the environmental impact of their activities. Indeed, they all agreed to take actions in order to have global warming not exceed 2°C.

Because industry in general, and moreover the fashion industry, are known as the most polluting activities, all emissions of carbon emitted along the production and distribution process accelerate the melting of polar ice caps and thus have a strong impact on rising sea levels. Typically, it concerns a lot of big countries, such as India and China, known worldwide as the largest fashion suppliers.

The goal is to develop a "low carbon economy." The Cop21 also forces companies on a worldwide scale to reach more sustainable processes concerning the production and culture of fabrics and yarn that engage a lot of water and energy resources (as we previously explained). Most countries chose to create carbon tax policies in order to achieve this goal. This adds to previous laws and policies each country had already put in place, such as the Ecotax in France and in Europe, or the Prevention Pollution Act in the United States[23].

6.6 Overview of African fashion and luxury: Showing the way!

Rarely is Africa considered as a trend leader. The African luxury and fashion industry has still not proved that it will be one of the leading markets in the global luxury industry. However, at this stage of globalization, Africa, "the smallest region for luxury goods, is expected to grow at 5.6% a year" (Aitken et al., 2015, p.1) and has ¼ of the raw materials used to manufacture international luxury items. Indeed, with the economic crisis of 2008, the potential for emerging market countries has risen, shading the Western mature markets. However, "as Asian economic growth slows, luxury goods companies may accelerate their presence in Africa to capture untapped demand" (Aitken et al., 2015).

However, although the consumption of luxury goods in mature markets, such as Europe and the United States, has declined, African market designers still struggle to be recognized in the global marketplace even if today African print is increasingly present in mainstream fashion. In 2012, the International Herald Tribune organized "The Promise of Africa," which is a conference that aimed to show the power of African fashion and to boost its resonance by challenging the stereotypes with which the continent has long been associated.

Nowadays international luxury brands and figures have acknowledged that Africa has become a future provider for raw material and thus a future partner in building sustainable luxury and fashion with the unique know-how and traditional skills of the African people. Indeed, Kim Jones, director of menswear at Louis Vuitton, said

[23] https://www.epa.gov/laws-regulations/summary-pollution-prevention-act

in 2012 that "Africa has immaculate craftsmanship, which lends itself to luxury" (Doran, 2014). Sustainability has been an element of brand differentiation and a source of long-term competitive advantage (Ricchetti and Frisa, 2011). Moreover, the growing demand for sustainable fashion pushes companies to review their business models, developing strategic partnerships and new relationship modes with all of the actors involved, in a sort of "green agreement" based on cooperation and sharing objectives. Specifically, the new business models are based on investments in sustainability, savings in terms of resources used (water and energy in particular), less waste of materials, reduction of cost of unsustainability (deriving, for example from legal impositions), technological innovations that translate into the ability to introduce new products, adoption of new collaborative practices among the actors of the supply chain, and greater attention to the relationships with both the local communities and the customers.

In the creation of the collections, however, the choice of material is not always the first step. While some African designers are inspired by a direct contact with the textiles, others perform effective brainstorming to choose the materials that would best suit the themes they wish to explore, their actual designs, or more simply their brand image and targets. Some African designers explain the practical choice to import Super 100 cotton textile, *"I essentially work with the Super 100 cotton because it is more flexible and can be worn everywhere unlike the Bazin which is considered heavy and cumbersome. People wear Bazin especially on Friday while tradi-modern outfits can be worn on a daily basis."* Therefore, due to a weak internal market for raw materials, African designers are left with no other option but importation. All of them, based in Africa or not, acknowledge the huge potential that lies within the continent in terms of unlimited natural resources. Designers consider the availability of raw materials as African designers' best asset because it extends the horizon of their creativity and allows them to create unique pieces from materials that are rare. For other African luxury and fashion brands, the opportunity of Africa is that it would give them a wide choice of textiles and the opportunity to hire more tailors to produce creations, and consequently more time to focus on the designs.

Despite a real will to be the future provider of Western countries, the industry leaders control Africa's natural resources because they fix the prices of the raw materials, transform them, and resell them at a high price. Take the example of leather, which is exported to Europe because West Africa does not have good tanning facilities and resold in Africa at a very high price. The West is also blamed for the destruction of the cotton industry that, in its opinion, is composed of multiple niches that can be factors of development. Indeed, the cotton industry encompasses agriculture, ginning, spinning, weaving, and dyeing, which are all potential areas of employment.

6.7 Conclusion and challenges

Some 35 years ago, one of us was lucky enough to be able to work as a research chemist in Switzerland. He was taught how to include the waste treatments price within the global manufacturing costs… and fish came back to live in the Rhine. Today, the challenges are numerous, but all of them strongly depend on mindsets and forecasted societal shifts. The worldwide fertility rate growth is known and the roadmaps are very few. Fashion and luxury are the two rare industries in which upgraded retail prices could be allowing the "real green" brands to be getting their return on investment as early as possible, and they definitely do not need the mass market ones to be cannibalizing this wonderful branding facet, which could therefore be making them understood as old fashioned, unfair, or… even kitsch. Ideally due to the high environmental impact by both the luxury and fashion brands, it is important to consistently and holistically aim for a sustainable business model which also would make the consumers adapt a sustainable consumption lifestyle.

References

Aitken, D., Rakic, M., Baldeira, S., 2015. Africa luxury goods market: full of untapped promise. https://www.bloomberg.com/professional/blog/africa-luxury-goods-market-full-of-untapped-promise/.

Cassaniti, M., 2015–2016. Luxury Consumption & Sustainability. http://tesi.eprints.luiss.it/16504/1/659241.pdf.

Davies, I.A., Lee, Z., Ahonkhai, I., 2012. Do consumers care about ethical-luxury? J. Bus. Ethics 106, 37–51.

Doran, S., 2014. Sustainable South African Luxury: Hanneli Rupert, Founder, Okapi. https://www.luxurysociety.com/en/articles/2014/01/sustainable-south-african-luxury-hanneli-rupert-founder-okapi/.

Fernandes, P., 2014. https://noisegoods.com/blogs/underthesun/14880613-why-vegetable-tanned-leather.

Freitas, A., Zhang, G., Mathews, R., 2017. Water Footprint Assessment of Polyester and Viscose and Comparison to Cotton. http://waterfootprint.org/media/downloads/WFA_Polyester_and__Viscose_2017.pdf.

Keinan, A., Crener, S., 2015. Stella McCartney. Harvard Business School Case 515-075, January (Revised November 2016).

Kering, 2018. http://www.kering.com.

Lochard, C., Murat, A., 2011. Luxe et développement durable: La nouvelle alliance (Eyrolles-Éd). d'Organisation, Paris.

Ricchetti, M., Frisa, M.L., 2011. Il bello e il buono: Le ragioni della moda sostenibile. Marsilio, Venezia.

Further reading

Chhabra, E., 2016. The Fabric of Our Lives or the Planet's Latest Threat? Fashion Startups Look Into Cotton Alternatives. https://www.vogue.com/article/fashion-startups-cotton-alternatives.

Diallo, M.E., Manière, I.C., 2015. The Global Struggle of African Fashion and Luxury. Thèse.

Hendriksz, V., 2017. Boohoo & TK Maxx among online retailers selling real fur labelled as fake. https://fashionunited.uk/news/fashion/boohoo-tk-maxx-among-online-retailers-selling-realfur-labelled-as-fake/2017122027362?utm_source=FashionUnited+UK+Trade+Journal&utm_campaign=94ec8d0eba-EMAIL_CAMPAIGN_2017_12_21&utm_medium=email&utm_term=0_18295b0d8e-94ec8d0eba-149669469.

Hymann, Y., 2016. Material Guide: Is Bamboo Fabric Sustainable? https://goodonyou.eco/bamboo-fabric-sustainable/.

Mekonnen, M., Hoekstra, A.Y., 2011. The green, blue and grey water footprint of crops and derived crop products. Hydrol. Earth Syst. Sci. 15, 1577–1600. https://www.hydrol-earth-syst-sci.net/15/1577/2011/https://doi.org/10.5194/hess-15-1577-2011.

Muthu, S.S. (Ed.), 2018a. Environmental Footprints and Eco-design of Products and Processes. SGS Hong Kong Limited, Hong Kong.

Muthu, S.S. (Ed.), 2018b. Models for Sustainable Framework in Luxury Fashion. Textile Science and Clothing Technology, Springer.

Noble, B., 2017. Fashion: The Thirsty Industry. https://goodonyou.eco/fashion-and-water-the-thirsty-industry/.

Oberhuber, N., 2014. Grün, grün, grün sind alle meine Kleider. http://www.faz.net/aktuell/finanzen/meine-finanzen/geld-ausgeben/oekomode-liegt-voll-im-trend-dieangesagtesten-label-12933625.html.

Smit, L., 2017. How Textile Industry Reduces its Water Footprint. https://www.ispo.com/en/trends/id_79705746/how-textile-industry-reduces-its-water-footprint.html.

Visser, M., Gattol, V., van der Helm, R., 2015. Communicating sustainable shoes to mainstream consumers: The impact of advertisement design on buying intention. Sustainability 2015 (7), 8420–8436. https://doi.org/10.3390/su7078420.

Wagner, E., Mark-Herbert, C., 2016. Relationship marketing in green fashion—a case study of hessnatur. In: Muthu, S.S., Gardetti, M.A. (Eds.), Green Fashion, Environmental Footprints and Eco-Design of Products and Processes. https://doi.org/10.1007/978-981-10-0245-8_2.

Analysis of water consumption and potential savings in a cotton textile dye house in Denizli, Turkey

Fatma Filiz Yıldırım*, Barış Hasçelik[†], Şaban Yumru*, Sema Palamutcu[†]
*Ozanteks Tekstil San ve Tic A.S R&D Center, Denizli, Turkey, [†]Pamukkale University, Engineering Faculty Textile Engineering Department, Denizli, Turkey

7.1 Introduction

The estimated fresh water resource in the world is only 3% of the world's total water reserves, which are underground water or water in the solid phase at the poles. Only 1% (Alper, 2015; Dvarioniene and Stasiskiene, 2007) of fresh water sources have been accessed so far. Water is vital for humankind and the rest of the living organisms. Nowadays in the world, 2.6 billion people live without hygienic conditions and 884 million people live far from clean water resources (Alper, 2015; Dvarioniene and Stasiskiene, 2007). Today, one in six of the world population does not have a reliable source of clean water to meet daily cleaning requirements. Currently, water demand is about 4500 billion cubic meters (Alper, 2015; Dvarioniene and Stasiskiene, 2007), and it is estimated that this amount will be 6900 billion cubic meters (Alper, 2015; Dvarioniene and Stasiskiene, 2007) in 2030. Due to industrialization, clean water is polluted and already limited fresh water resources have been reduced. According to research, many countries nowadays have only half of the amount of water they had in 1975, and water demand will double in the near future (Alper, 2015; Dvarioniene and Stasiskiene, 2007).

Turkey is neither a water-rich nor a water-poor country in which existing water resources should be consumed carefully and not be polluted; a water consumption forecast should be completed both for social and industrial purposes (Alper, 2015). Nowadays, water resource management is becoming an important environmental issue. The increased cost of reliable fresh water supplies and wastewater recycling requires intensive technological applications that are more environmentally friendly (Alkaya and Demirer, 2014).

When the 2014 world textile and clothing exports data are analyzed, it may be seen that Turkey is ranked seventh with a 12.6% share of the total world market (Shaikh, 2009)[1]. The Turkish textile and clothing sector has Europe's largest production capacity of yarn, home textile, and denim fabric manufacturing capacities. The denim fabric

---

[1] https://ekonomi.isbank.com.tr/UserFiles/pdf/sr02_tekstilsektoru.pdf

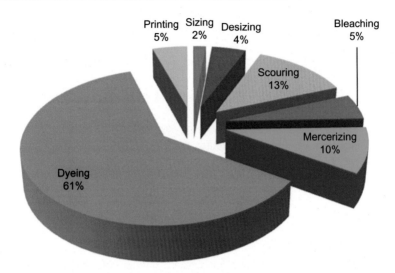

Fig. 7.1 Average water consumption ratio for cotton wet operations (N.C. Division of Pollution Prevention and Environmental Assistance (NCDENR), 2009).

export rank of Turkey is reported as third in the world market and the towel export rank of Turkey is reported as fourth in the world market (Shaikh, 2009)[1].

In the textile industry, a large amount of water is consumed in various processing operations. The Textile dye house processes are one of the major fresh water-consuming processes in the textile manufacturing process chain. Water consumption in the spinning and weaving processes are nearly negligible compared with the textile wet processes of pretreatment, dyeing, printing, and finishing operations (N.C. Division of Pollution Prevention and Environmental Assistance (NCDENR), 2009; Defrawy, 2002)[2]. Average fresh water consumption ratio values for conventional cotton textile manufacturing operations are given in Fig. 7.1. The highest water consuming process is reported as the dyeing process with a water consumption ratio of 61% and the lowest ratio is reported as the sizing operation at weaving preparation with a ratio of 2% (N.C. Division of Pollution Prevention and Environmental Assistance (NCDENR), 2009).

In textile wet processing applications, dyes, auxiliaries, additives, and finishing chemicals are applied to the fabric in an aqueous environment via bath to comply with the required processing steps of sizing, scouring, desizing, bleaching, mercerizing, dyeing, and finishing. In textile wet processes, water is not only used as a solvent for chemicals, but also for ion exchange, cooling water, streaming, and cleaning processes (N.C. Division of Pollution Prevention and Environmental Assistance (NCDENR), 2009).

In the past, water resources were abundant and inexpensive. Additionally, textile wastewater could be discharged to the environment and this discharge did not cause

---

[2] https://www.iski.istanbul/web/tr-TR/musteri-hizmetleri/su-birim-fiyatlari

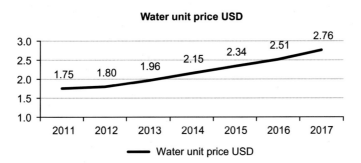

Fig. 7.2 Changes of industrial fresh water costs in Turkey, USD/tons (Environmental Technology Best Practice Programme (ETBPP), 1997).

excessive costs or have any restrictions (Alkaya and Demirer, 2014). Today, the cost of fresh water is higher than before. The increasing cost of fresh water and the additional cost for the disposal of textile wastewater are among the triggering factors for being cautious on the water consumption levels of traditional textile dye houses. Governmental rules and rising environmental actions in the world have pushed the dye house managements to improve and install more environmentally friendly technologies in their plants (Alkaya and Demirer, 2014). The increasing cost of industrial fresh water over time is given in Fig. 7.2.

The textile industry has been forced to make reductions of unit water consumption per kg of product intensively since the end of 20th century starting with the Western countries. Increased competition for clean water continues as a result of limited fresh water resources and the rising demands from both industrial and residential growth in the world (Shaikh, 2009). Various water saving and pollution prevention models have been considered and applied in textile processes. A water management model, the integrated water resource management (IWRM), is offered as a useful tool for complex water research using production systems in industrial companies. The Best Available Techniques (BAT) are presented as another program to the textile wet processing sector offering reduction or elimination of wastewater discharge and emissions. The European Directive, "Integrated Pollution Prevention and Control (IPPC)" will be decisive in sustainability and encouraging water recycling and reuse applications (Alkaya and Demirer, 2014). In December 2011, "Communiqué Integrated Pollution Prevention and Control in the Textile Sector" was put into effect by the Ministry of Environment within the scope of the EU Acquisition Program (N.C. Division of Pollution Prevention and Environmental Assistance (NCDENR), 2009).

In this chapter, a general literature review of water and energy utilization analysis studies, reduction strategies, and models is given. A Pareto analysis of process optimization in the dye house is introduced that explains the reasons for the replacement of the overflow rinsing process in the HT rope dyeing machine with the neutral enzyme utilized drop-fill process. A case study of the reduction of water, chemical, and energy consumption to provide a reduction in the environmental impact of a large textile mill in the city of Denizli, Turkey is given in detail.

7.2 Literature review

Sustainable production and optimization of water, chemical, and energy approaches has been successfully accomplished in many textile plants in different countries around the world. In these applications, various techniques and technologies have been explored to reduce the chemical, water, and energy demands of the companies[1].

As a result of systematic, intensive, constructional, and even education/training-based optimization work and research, the amounts of consumed chemical auxiliaries, fresh water, and associated textile wastewater have been reduced. Some general findings to reduce textile wastewater discharges are listed below (Ozturk et al., 2009; Ferrero et al., 2011; Oner and Sahinbaskan, 2011)[1]

- In sizing process: use of low Biochemical Oxygen Demand (BOD) size agents.
- In desizing process: use enzymes as de-sizing agents.
- In scouring process: employ solvent-aided processes.
- In bleaching process: use of hydrogen peroxide or ammonium salts instead of chlorine-based chemicals
- In mercerizing process: install and utilize a caustic soda recovery unit.
- In dyeing process: implement pad-batch systems; use low-sulfide dyes instead of sulfur-containing dyes; substitute some auxiliary materials with ethanol (Ozturk et al., 2009; Ferrero et al., 2011; Oner and Sahinbaskan, 2011)[1].
- In finishing process: avoid use of preservation agents (Ozturk et al., 2009)[1].

In sizing, a new process was developed for combined pretreatment and dyeing. This process allowed decreasing the process time and increasing the water use efficiency. This process is called Rapid Enzymatic Single-bath Treatment (REST). This method has been reported to yield a 50% saving in time and a 66% saving in water compared to the conventional process (Tanapongpipat et al., 2008)[1]. Other research involves the scouring process, in which the process was optimized by changing the concentration of desizing agents, dipping time, and flotte temperature. With this optimization, 89% of the sizing agent was eliminated without any performance loss in the process and intermediate products (Tanapongpipat et al., 2008)[1]. Another experimental research effort concerns the reuse potential of textile dye house wash water effluents, in which 64% of consumed plant water is reused efficiently (Jiang et al., 2010)[1]. By optimizing production schedules, it is possible to reduce production time, the amount of wastewater discharged, and the fresh water consumption. To achieve this aim, research groups developed a genetic algorithm and implemented it in a textile mill. The results show that the water consumption values were reduced by 20%–30%, the amount of wastewater was reduced about 20%, and the production time was reduced 10%–15% (Thiede et al., 2013)[1].

In Lithuania, an Integrated Water Resource Management (IWRM) model was used for evaluating water consumption in the process industry. The IWRM model is designated as a useful tool for systematically assessing methods for the different processing industries to reduce fresh water use and improve wastewater reuse possibilities. By applying this model to different industrial processes, there are possibilities to create various scenarios in order to optimize the management of water resources in single-source production processes or multiprocess systems of integrated continuous production companies. The result of this is that a small textile mill, applying the IWRM, has managed to save 62% water in the designated approach (Alkaya and Demirer, 2014).

In Turkey, an Environmental Performance Evaluation study was performed in a woven fabric manufacturing mill in Bursa. With this project, the environmental performance of the company was evaluated and water, energy, and chemical consumption were reduced[1]. The results that were achieved on the environmental impact as a result of the applied project are summarized in Table 7.1 and Fig. 7.3.

As a consequence of these applications, the total water consumption of the company decreased by 40.2% while the amount of wastewater discharged decreased by about 43.4%[1]. In another industrial evaluation project, six different models were proposed for sustainable and green production transformation in textile companies. With process optimization and/or some minor revisions, water and chemical auxiliary consumption are reduced, electric and natural gas consuming processes and areas are

Table 7.1 Summary of environmental impact before and after applications[1]

Resources	Specific consumption values		
	Before application	After application	Change (%)
Water consumption (l/kg fabric)	138.9	83	−40.2
Wastewater (l/kg fabric)	124.1	70.2	−43.4
Salt consumption (kg/ton)	218.7	118.2	−46.0
Energy consumption (kWh/ton fabric)	8466.0	7021.0	−17.1

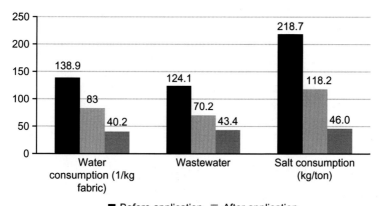

Fig. 7.3 Water, salt consumption, and wastewater reduction in the Environmental Performance Evaluation model in the company.

determined; then cleaner production possibilities, eco-efficiency potentials, and improved production approaches are indicated for implementation[3].

In Bangladesh, recommended best practices for in Bangladeshi Factories were developed with the motivation of the Natural Resources Defense Council (NRDC) and World Bank. The recommended water and energy saving practices that were applied are: (1) elimination of water leaks by reducing hose pipe usage; (2) reuse of cooling water in dyeing machines; (3) reuse of rinsing water as process water; (4) application of steam system management; (5) insulation of steam pipes and valves; and (6) recovery installation to maintain heat from the drying operations[4].

In addition to chemical and water use efficiency, energy efficiency has attracted a number of academics[1]. They found that a seven-stage approach developed for the industry provides a significant reduction in energy consumption (Hasanbeigi, 2010)[1]. The steps are given below:

- Energy portfolio;
- Macro-analysis;
- Measurement;
- Modeling/analysis;
- Identification;
- Evaluation;
- Implementation (Hasanbeigi, 2010)[1].

Implementation of the recommended energy management approach provides the possibility of up to 6% energy reduction at companies (Hasanbeigi, 2010)[1]. In other approaches, new dyeing processes, such as the supercritical dyeing technique, ultrasonic-assisted dyeing, and foam technology, have also been proposed in several studies (Kar et al., 2012)[1].

7.3 Method

The company manufactures different kinds of products, such as different weights of woven material and weft-knitted and warp-knitted fabrics to manufacture bathrobes, towels, bed linens, cushions, coverlets, seamless underwear, and sportswear. The total manufacturing area of the plant is 200,000 m^2 and employment is 1400 blue- and white-collar workers in total. Manufacturing of fabric is carried out continuously by a total of 300 terry fabric and simple fabric weaving looms and 100 underwear circular knitting machines and weft knitting machines. Wet processing is carried out continually by 21 rope dyeing machines (HT), 7 polyamide dyeing machines, and 5 cone dyeing machines. Also, to manufacture garments, to make side sewing, and to complete the ultimate products, 700 sewing machines are run together. The annual production of towels is 30 million pieces, for bathrobes 1.8 million, for bed linen 1 million pieces, and seamless garment production is 3.6 million pieces. The company has OEKO-Tex 100, IMO organic, TSE, BSCI, Sedex, GSV, and ISO 27001 standards.

---

[3] http://www.mirachemindustries.com/products/Enzymes/133-EMCOZYME-CRS.pdf
[4] http://www.lebanese-cpc.net/CP_Textile.pdf

Additionally, the company has solar power installations in fields and on factory roofs that has an installed capacity of 17 MW.

Optimization of water consumption in operations is an important issue due to the huge consumption of fresh water. The water management project within the company is designed to prevent unnecessary wastewater discharge and to collect, reuse, and reduce the plant water during the production stages.

The Pareto analysis technique is used to define the problem and to designate the possible choices to reduce water consumption at the plant. As is known, Pareto analysis is a decision-making technique that separates a statistically limited number of input factors into the largest effect on a desired or undesired result. Pareto analysis is based on the idea that 80% of a benefit can be achieved by carrying out 20% of the work, or 80% of the problems based on 20% of the causes[5]. The Pareto analysis system determines which departments will create the most problems in the organizations or sectors.

7.3.1 Analysis of overflow rinsing process

Therefore, in this study, a Pareto analysis has been performed for the annual amounts of water consumption in all departments of the plant and the results are given in Fig. 7.4. As seen in Fig. 7.4, the dye house is the most water-consuming department with 49,745 tons/year and 92.3% of the water consumed at the plant.

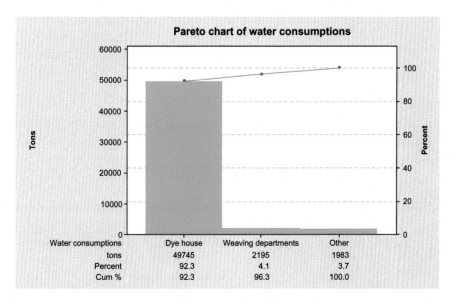

Fig. 7.4 Average annual water consumption amounts of the manufacturing departments in the plant.

---

[5] https://www.investopedia.com/terms/p/pareto-analysis.asp

After the most water-consuming department is determined, water, chemical, and energy consumption values should be evaluated. In the dye house department of the plant, all wet processing processes (HT dyeing, cone dyeing, pad patch dyeing, polyamide dyeing, other) are analyzed individually using the Pareto method. The results of the Pareto analysis in the Minitab evaluation program showed that the HT (high temperature) dyeing department was the highest water-consuming department in the dye house and the cone dyeing department was ranked second after the HT department. The Pareto chart of each dye house operation in the plant is given in Fig. 7.5. On average, a total of 50,000 tons of water is consumed and about 66.4% of the consumed water is utilized in the HT dyeing department, while 20% of the total water consumption in the plant is consumed in the cone dyeing process.

As result of the Pareto analysis, the HT dyeing department and the cone dyeing department were selected for the water management project.

In the close analysis work that was carried out for both the HT rope dyeing and cone dyeing processes, the amount of water consumption in each process step was analyzed for light shade coloration of cotton fabric and yarn with reactive dye. The process steps are shown for each process in terms of water consumption, time requirements, and process temperature in Fig. 7.6 for the HT rope dyeing machine and in Fig. 7.7 for the cone dyeing machine.

As known from the literature, most of the water consumption in reactive dyeing occurs in overflow rinsing and its share of the water consumption is nearly 69%[3]. To find the water consumption ratio at each step of the HT rope dyeing process, consecutive Pareto analyses were performed to determine the most water-consuming step

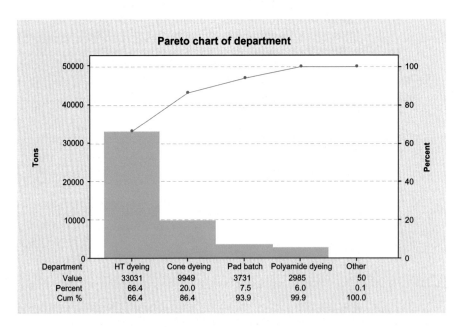

Fig. 7.5 Average annual water consumption ratios of the dye house department of the company.

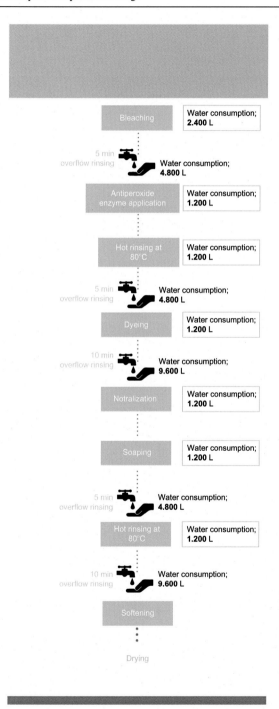

Fig. 7.6 Reactive dyeing process for light shades in HT department.

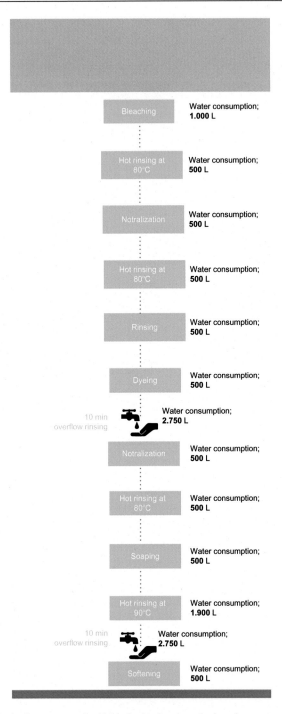

Fig. 7.7 Reactive dyeing process for light shades in cone dyeing department.

of the reactive dyeing in the HT rope dyeing machine. HT reactive dyeing water consumption values are given in Table 7.2 and the applied Pareto analysis results are given in Fig. 7.8.

Water consumption values for the cone reactive dyeing process are given in Table 7.3 and the applied Pareto analysis results are given in Fig. 7.9.

As a result of the Pareto analysis of the HT rope dyeing and cone dyeing machines, it was found, parallel to the literature, that rinsing steps have a 76% share of the total water consumption in HT reactive dyeing and 46.2% of total water consumption in the cone reactive dyeing process.

Table 7.2 Amount of water consumption in HT reactive dyeing

HT reactive dyeing process	Amount of water consumption (L)
Bleaching	4800
Dyeing	3600
Rinsing	2400
Overflow rinsing	33,600

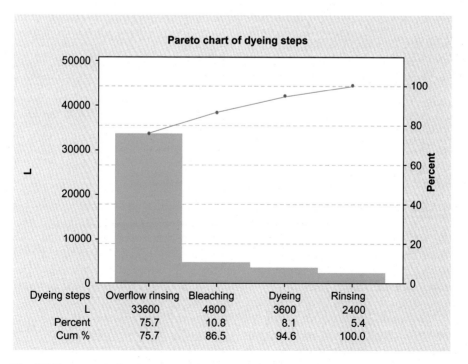

Fig. 7.8 Pareto analysis of water consumption in HT reactive dyeing process.

Table 7.3 Amount of water consumption in cone reactive dyeing

Cone reactive dyeing process	Amount of water consumption (L)
Bleaching	1500
Dyeing	1000
Rinsing	3900
Overflow rinsing	5500

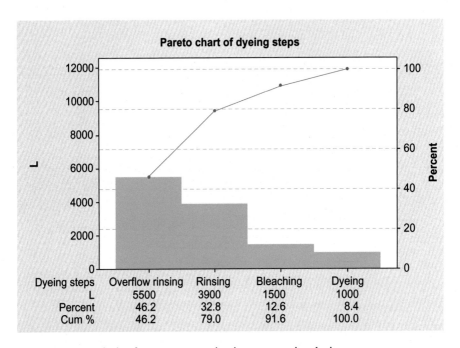

Fig. 7.9 Pareto analysis of water consumption in cone reactive dyeing process.

7.3.2 Implemented of the drop-fill process

As a result of the Pareto analysis, it was decided to research new alternative methods to decrease water consumption. As an alternative method, the drop-fill process was selected to replace the overflow rinsing process. Machines were installed and processes reorganized to implement the drop-fill process and a water consumption analysis was completed alongside fabric quality and performance evaluations. The potential impact of the drop-fill process on product quality was evaluated in terms of the colorimetric and fastness values of the fabrics.

The drop-fill process uses enzymes to enhance the fabric dyeing operation. It is known that application of acidic enzymes in the predyeing or postdyeing steps helps to enhance the antipilling properties of the fabric. The literature indicates that a reduction in the number of dye baths has become possible with utilization of neutral enzymes (Balcı et al., 2010)[5]. Using neutral enzymes instead of acidic enzymes provides one dye bath benefit in the HT rope dyeing process steps.

Influence on the fabric quality parameters were investigated with the colorimetric and fastness properties of fabrics processed with neutral enzymes. Additionally, some experiments were carried out to remove the use of dispersant and leveling agents used for uniform distribution of dyes.

After a pilot application and evaluation of the fabric quality indicators, the overflow rinsing operation in all HT reactive dyeing processes was replaced incrementally with the drop-fill process using neutral enzymatic treatment.

The drop-fill process replaced three overflow rinsing processes in light color shades, four overflow rinsing processes in dark and medium color shades, and two overflow rinsing processes in white colors. Using neutral enzymes instead of acidic enzymes provided a dye-bath gain. Additionally, soap usage was reduced and the use of dispersant and leveling agents was abolished. The changes in HT rope dyeing are shown in Figs. 7.10 and 7.11, and changes in the cone dyeing processes are shown in Figs. 7.12 and 7.13.

In the HT dyeing process, water savings was observed in shifting from overflow rinsing to drop-fill rinsing: 50,000–15,000 L in dyeing of medium shades, 65,000–18,000 L in dyeing of dark shades, and 23,000–9000 L in the dyeing of white colors. The total savings in the water and chemical consumption rate is estimated to be about 70% for the HT reactive dyeing process. The highest reduction in water consumption was observed for dark color shades.

In the cone dyeing process, water consumption was decreased from 12,000 to 6000 L in the dyeing of light shades, 14,000–7000 L in the dyeing of medium shades, 16,000–7000 L in the dyeing of dark shades, and 4000–3000 L in the dyeing of white colors. The total savings in the water and chemical consumption rate was estimated to be about 54% for the cone reactive dyeing process. The highest reduction in water consumption was observed for dark color shades in cone dyeing.

After the implications of piloting the drop-fill process were known, the rest of the machinery was adapted to the drop-fill process and an efficient reduction in the water and energy consumption was maintained. The saved water and energy consumption rates are given Fig. 7.14. The daily average energy consumption in the HT dyeing department decreased from 10,089 kWh per day to 8546 kWh per day as a result of these applications. The specific energy consumption for the HT reactive dyeing process decreased by 15% on average. The daily average energy consumption of the cone dyeing department was decreased from 5136 kWh per day to 3827 kWh per day as a result of these applications. The specific energy consumption for the cone reactive dyeing process decreased by 26% on average.

7.4 General findings

The application of the shifted drop fill process has added additional advantages to the process. Process durations have been shortened and additional energy savings have been provided. Furthermore, as a result of the reduced water consumption, the salt

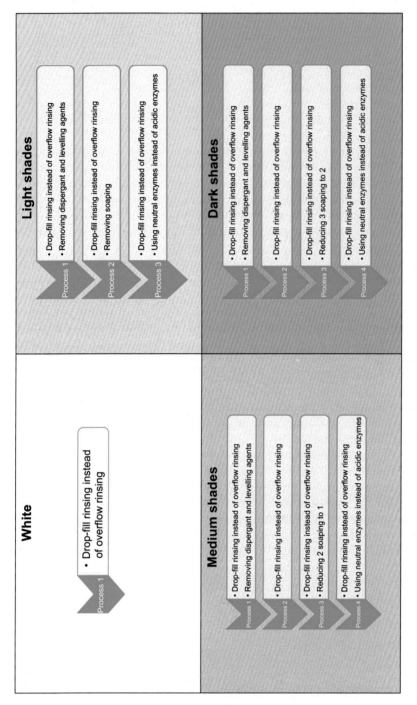

Fig. 7.10 Reduced water and chemical consumption with decreased number of process steps in HT reactive dyeing.

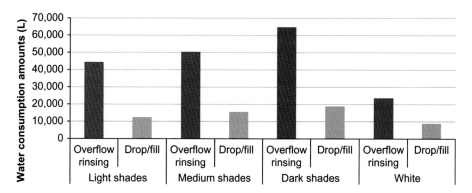

Fig. 7.11 Reduction in consumed water consumption in drop-fill process in the HT department process.

consumption has also been reduced. The reduction of salt consumption runs in parallel to the reduction of water consumption. For producing 1 ton of soft water, 3 kg of salt are used. The 62% total average water savings in the HT and cone dyeing departments provides approximately 20,000 tons of water savings and 60 tons of salt per month.

The total water consumption values in the dye house were remarkably reduced. These values are shown in Fig. 7.15. As seen in that figure, there are also some other significant decreases. These declines may be due to seasonal fluctuations in order and product types. However, the water consumption in the dye house declines remarkably. This can be seen more clearly by examining the amount of water used per kilogram of fabric (Fig. 7.16).

As seen in Fig. 7.16, the initial water consumption of reactive dyeing was about 120 L/kg of product. After shifting the process to the drop fill method, it was reduced to 46 L/kg of product.

The annual cost savings as a result of the implemented production application is given below in Table 7.4).

As result of the improved processes in the plant, a savings of 146,216 USD/year was achieved. Beside the money saved, the environmental impact of the dye house has also been decreased by ensuring efficient use of reduced water resources, reduced energy consumption, and reduced salt usage.

Worldwide, the annual cotton fiber grown for the season of 2016/17 is about 22 million metric tons and most of the cotton grown will be processed through and clothing manufacturing lines. Every kilogram of virgin cotton will be sent to the harsh water-consuming processing stages and household-based consumption stages that consume large amounts of fresh water. A rough estimation of 20 million metric tons of cotton fiber will create an annual 22 billion metric tons of water footprint during its journey from the cotton plantation to the waste disposal site[6].

---

[6] http://www.temizuretim.gov.tr/Files/haberfiles/d120215/Prof.%20Dr.%20Mehmet%20KANIK-Tekstil%20Terbiye%20Sekt%C3%B6r%C3%BCnde%20Temiz%20%C3%9Cretim%20%C4%B0%C3%A7in%20%C3%96rnek%20Modellerin%20Olu%C5%9Fturulmas%C4%B1.pdf

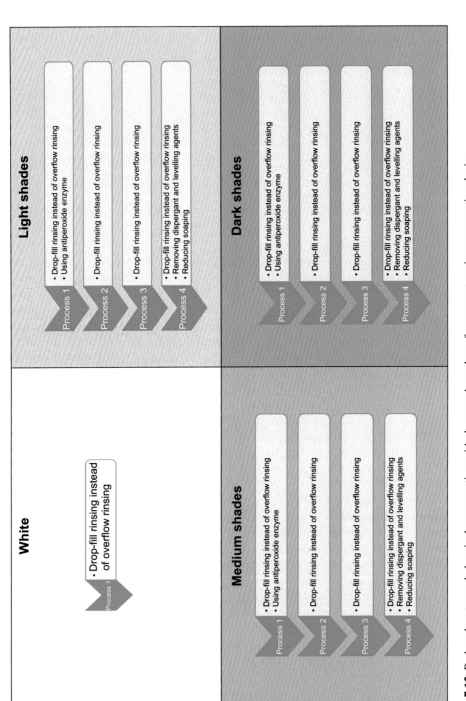

Fig. 7.12 Reduced water and chemical consumption with decreased number of process steps in cone reactive dyeing.

Analysis of water consumption and potential savings

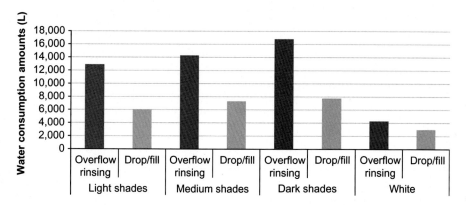

Fig. 7.13 Reduction in consumed water consumption in the drop-fill process in the cone department.

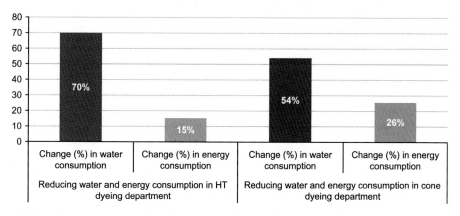

Fig. 7.14 Total water and energy consumption change rates in the HT and cone dyeing processes.

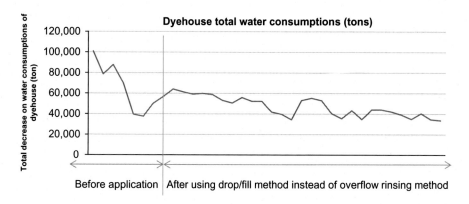

Fig. 7.15 General change of the water consumption amounts in the dyeing department.

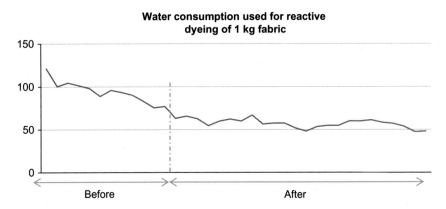

Fig. 7.16 General change in the water consumption in the dyeing department per kilogram of fabric.

Table 7.4 Cost savings as a result of the implemented production application

Cost items	Annual cost savings (USD/year)
Water	78,002.4
Salt	31,441.8
Energy	31,051.2
Chemical	5720.9
Total	146,216

In Turkey, the average value of annual cotton fiber consumption is about 1.4 million tons and all of it requires the intensive fresh water-consuming phases of scouring, mercerizing, dyeing, washing, and other wet finishing processes. In the wet process stages of a regular 250 g cotton t-shirt, the manufacturing phase requires about 35–40 L of plant water in a traditional manufacturing line. When it is accepted that all of the cotton fiber used in the Turkish industry is manufactured as a regular t-shirt, then it can be estimated that in total all of the 1.4 million tons of cotton fiber ends up as 1.1 million tons (4.4 million of 250 g) of t-shirts. Therefore, it can be stated that in total 154–176 million tons of plant water is consumed in the cotton sector[7].

In the case of this study, the plant water consumption was reduced from 120 L/kg to 46 L/kg of cotton fabric in which about 60% water saving has become possible. Such water saving initiatives and widened sectoral water saving or efficiency improvement programs would provide many fresh water saving possibilities to the wet processing sector.

---

[7] http://awsassets.wwftr.panda.org/downloads/su_ayak_izi_raporweb.pdf

7.5 Conclusion

In this study, water consumption, chemical consumption, energy consumption, and salt consumption values were examined for potential sustainable production process improvement applications. Primarily, water-consuming departments were investigated. Wet processes, especially the HT and cone dyeing departments, were found to be the highest water consuming processes in the plant. Then, with the literature survey, the best applications were investigated to achieve saving in water, chemicals, salt, and energy. After evaluating the appropriate water saving possibilities in the plant, the following applications were implemented:

- Use drop-fill rinsing instead of overflow rinsing.
- Use neutral enzymes instead of acidic enzymes.
- Reduced salt consumption.
- Reduced processing time and energy.

In consequence of the sustainable production process improvement work, the total average water consumption of the company was decreased by about 62% and the duration of the processes were shortened as well. Because energy is consumed primarily for heating the dyeing effluents, the reduced water consumption and processing time resulted in reduced energy consumption. The reduction in total average energy consumption was about 20.5% in total in the HT rope and cone dyeing departments.

The promising results that have been gathered as water, energy, salt, and time savings show that spreading these approaches would help to reduce the environmental influence of the traditional cotton dyeing process in general.

References

Alkaya, E., Demirer, G.N., 2014. Sustainable production: a case study from a woven fabric manufacturing mill in Turkey. J. Clean. Prod. 65, 595–603.

Alper, F., 2015. Sürdürülebilirlik Kavramı içerisinde Su Ayak İzi: Tekstil Sektörü Örneği. (Doctoral deissertation). Fen Bilimleri Enstitüsü.

Balcı, O., Asker, Ğ., Kurtoğlu, N., 2010. Biyoparlatma Ve Reaktif Boyama İşlemlerinin Kombine Uygulanması İle Hızlı Boyama Prosesi. Mühendislik Bilimleri Ve Tasarım Dergisi. 1(1).

Defrawy, N.E., 2002. Water management in industry: technical and economic aspects. Int. J. Environ. Stud. 59 (5), 573–587.

Dvarioniene, J., Stasiskiene, Z., 2007. Integrated water resource management model for process industry in Lithuania. J. Clean. Prod. 15 (10), 950–957.

Environmental Technology Best Practice Programme (ETBPP), 1997. Water and Chemical Use in Dyeing and Finishing (No. GG62). Retrieved from, http://www.wrap.org.uk/sites/files/wrap/GG062.pdf.

Ferrero, F., Periolatto, M., Rovero, G., Giansetti, M., 2011. Alcohol-assisted dyeing processes: a chemical substitution study. J. Clean. Prod. 19, 1377e1384.

Hasanbeigi, A., 2010. Energy-Efficiency Improvement Opportunities for the Industry. China Energy Group, Energy Analysis Department Environmental Energy Technologies Division Lawrence Berkeley National Laboratory.

Jiang, W., Yuan, Z., Bi, J., Sun, L., 2010. Conserving water by optimizing production schedules in the dyeing industry. J. Clean. Prod. 18, 1696e1702.

Kar, A., Keane, S.E., Greer, L., 2012. Best Practices for mills to Save Money and Reduce Pollution in Bangladesh. *Natural Resources Defense Council (NRDC)*, Dhaka, p. 28.

N.C. Division of Pollution Prevention and Environmental Assistance (NCDENR), 2009. Water Efficiency Industry Specific Processes: s. North Carolina. Retrieved from, http://infohouse.p2ric.org/ref/01/0069203.pdf.

Oner, E., Sahinbaskan, B.Y., 2011. A new process of combined pretreatment and dyeing: REST. J. Clean. Prod. 19, 1668e1675.

Ozturk, E., Yetis, U., Dilek, F.B., Demirer, G.N., 2009. A chemical substitution study for a wet processing mill in Turkey. J. Clean. Prod. 17, 239e247.

Shaikh, M.A., 2009. Water conservation in industry. Pakistan Journal 2009, 48–51.

Tanapongpipat, A., Khamman, C., Pruksathorm, K., Hunsom, M., 2008. Process modification in the scouring process of industry. J. Clean. Prod. 16, 152e158.

Thiede, S., Posselt, G., Herrmann, C., 2013. SME appropriate concept for continuously improving the energy and resource efficiency in manufacturing companies. CIRP J. Manufacturing Sci. Technol. 6, 204e211.

Further reading

Hoque, A., Clarke, A., 2013. Greening of industries in Bangladesh: pollution prevention practices. J. Clean. Prod. 51, 47e56.

Ulson de Souza, A.A., Melo, A.R., Pessoa, F.L.P., de Souza, G.U., de Arruda, S.M., 2010. The modified water source diagram method applied to reuse of industry continuous washing water. Resour. Conserv. Recy. 54, 1405e1411.

Water conservation in textile wet processing

8

M. Gopalakrishnan, V. Punitha, D. Saravanan
Department of Textile Technology, Bannari Amman Institute of Technology, Sathyamangalam, India

8.1 Introduction

Water is a basic necessity for treating textile materials in the wet processing industry. Due to increases in the standard of living, the use of textile materials has also increased a lot, which in turn necessitates an increase in the production of textiles and also an increase in the demand for water. Conventional machines and treatment methods employ huge amounts of water, especially in the case of natural fibers for which the demand is as high as 150 kg of water (van der Walt and van Rensburg, 2009) for every kilogram of material. Understanding the pressure arising from every part of the society to reduce water use by various industries, including textiles, newer processes and techniques have been developed by the machine as well as the dyes and auxiliaries manufacturers. The techniques for conserving water while processing textile products are listed here.

Techniques

- Low wet pickup techniques.
- Foam finishing.
- Cold-pad-patch method.
- Counter-current washing technique.
- Latest dyeing equipment.
- Solvent dyeing using supercritical carbon dioxide.

8.2 Low wet pickup techniques

The application of chemical finishes on a padding machine yields 65%–75% (Leah, 1982) wet pickup, which not only increases the water consumption but also the amount of energy required to dry the fabrics. Ever increasing oil and gas prices continuously escalate the cost of energy day by day. However, it is very difficult to reduce the wet pickup further by means of the mechanical action of squeezing. The squeezing operation removes only the surface liquor rather than the liquor in the interstitial section and in fiber capillaries. For cotton fabric, the lower limit of wet pickup is around 40%–45% (Leah, 1982) by means of mechanical action. Further reduction of wet pickup by means of squeezing is practically impossible. The rubber hardness, type of fiber, nip pressure, roller diameter, and speed of the machine decides the amount of wet pickup (Goldstein and Smith, 1980).

The water imbibition values are different for different fibers and fabrics and are partially related to the low wet pickup of different fibers. For example, the imbibition value for cotton is around 30%, whereas it is 3%–20% for most synthetic fibers (van der Walt and van Rensburg, 1984). Reducing the wet pickup further from the value of imbibition in normal padding machines is practically sought. In the application of finishes, liquor acts as a distribution channel for the chemicals. Reducing the wet pickup could possibly result in uneven distribution of chemicals and dyes because the capillary action may not function below certain limits. The wet pickup of the low add-on techniques lies in between 20% and 40%, whereas for the conventional padding method of application, it is around 70%–100% (Leah, 1982).

Low add-on techniques are achieved by means of two systems, namely, expression and topical techniques. In expression techniques, the fabrics are allowed to saturate in the liquor by immersing the fabric in the liquor and then removing the surplus liquor by squeezing or by means of vacuum extractors or dehydration systems or air-jet ejectors. In topical techniques, a precise quantity of liquor is applied on the fabric by the transfer method either by directly or indirectly using techniques that include kiss roller coating, loop transfer coating, wick system, spray system, engraved roller coating, and foam finishing (Leah, 1982).

8.2.1 Expression technique

8.2.1.1 Fiber-filled rollers

In the existing mangles, rubber-covered pad rollers are replaced with porous materials as illustrated in the Fig. 8.1. The top rubber roller is covered with coated fibers that can absorb liquor or water during squeezing and transport it through fine pores. The bottom steel roll in the existing pad is also replaced with a coated fiber bunch. According to a report, two fiber-filled roller arrangements can reduce the wet pickup up to 5% for knitted nylon fabrics (van der Walt and van Rensburg, 2009).

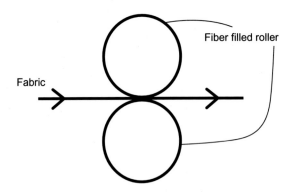

Fig. 8.1 Fiber-filled roller.

8.2.1.2 The dehydration system

Dehydration systems also reduce the wet pickup considerably during the treatment of textile materials. Although this system was adopted in industry many years ago, it has recently been found that this system has the potential to reduce the wet pickup considerably compared to conventional systems. The machine *Hydrofuga* was developed by Kleinewefers using this principle. The processing involves two endless nonwoven blankets, covered separately, in place of the top and bottom pad rollers used in the conventional padding, as shown in the Fig. 8.2. The pad rollers squeeze the fabrics along with the nonwoven blankets. The nonwoven blankets absorb the excess liquor from the fabric and dehydrate mechanically (van der Walt and van Rensburg, 2009).

8.2.1.3 Vacuum extractors

A vacuum slot extractor (Fig. 8.3) is the most common method used to reduce the wet pickup of padded fabrics. In this method, the fabrics are padded using the conventional padding mangle and then the fabric is passed over the vacuum slot extractor for removal of excess liquor. The lowest achievable wet pickup on the vacuum extractor for 100% polyester knitted fabric is around 10%–20% (Broadbent and Quebec, 1990). Different patterns of vacuum slot are available, but the straight and herring bone types are most commonly used. In order to achieve the maximum extraction, it is necessary that the fabric must be continuously in contact with the vacuum slot(s).

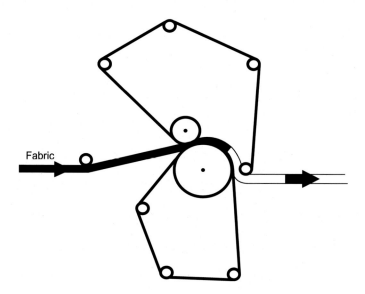

Fig. 8.2 Dehydration system of coating.

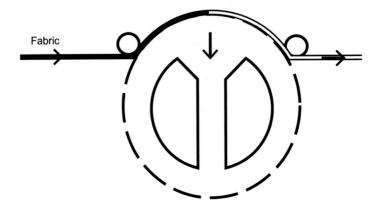

Fig. 8.3 Vacuum slot extractor.

8.2.1.4 Air jet ejectors

In air jet ejectors, compressed air is used to remove the excess water in place of vacuum (used in vacuum slot extractors). This technique is suitable for a wide range of fabrics and especially heavy fabrics, while lightweight materials and knitted fabrics are easily distorted.

In an air jet ejector (Fig. 8.4), the compressed air at high pressure is passed over the fabric in a slit, which is adjusted using the plates according to the thickness of the fabrics treated in the machine. Previously padded fabrics are passed over a pair of guide rollers under tension and the nozzle(s) is placed above the fabrics, in between the two guide rollers. The nozzle ejects the steam or air at high pressure and the excess moisture on the fabrics is taken away by the steam. The lowest wet pickup achieved by this process is around 10%–15% on synthetic fibers, while it is closer to 40%–50% for cotton and its blends (van der Walt and van Rensburg, 1984).

8.2.2 Topical techniques

In the expression method, the excess liquor is squeezed out from the fabric. Whereas, in topical methods, the precise amount of liquor is applied on the fabric either by direct application or indirect application methods.

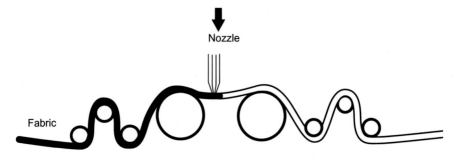

Fig. 8.4 Air jet ejector.

8.2.2.1 Engraved roller

In this engraved roller system (Fig. 8.5), the chemicals are applied on fabrics by means of an engraved roller. The basic setup of the engraved roll applicator consists of the top rubber roller, the bottom engraved roller, a doctor blade, and a trough. The engraved roller takes up the chemicals from the trough and applies them onto the fabric in the precise amount needed. The amount of chemicals transferred per unit area is based on the design and construction of the engraved rollers, the number of cavities, and the surface of the engraved rollers (Ashish Kumar, 2009).

The main disadvantage of this system is that the wet pickup is not uniform for different fabrics. It varies with fabric areal density, viscosity of the process liquid, engraved rollers, and the profile of the engraving used to achieve the desired wet pickup. A wet pickup of 10% is achievable on cotton fabrics, however, wet pickup lower than 20% significantly deteriorates the physical properties (van der Walt and van Rensburg, 2009). A rubber roller is used to remove the excess liquid instead of an oscillating doctor blade, as a modification developed by Sando Iron Works in YS-Mangle. Further, this system employs an uneven surface roller with small holes on the surface (Sando and Shidoshiro, 1981).

8.2.2.2 Kiss or lick roller coating

Kiss or lick rollers (Fig. 8.6) are used to coat the chemicals for special purposes for textile applications. For example, stiffening agents are applied on one side of corduroy fabric and book-binding cloth and chemicals are applied on cotton fabrics for crease recovery finish. Different kiss roll arrangements are used in the textile industry (Schindler and Hauser, 2004). Some of the arrangements are given in Fig. 8.6.

8.2.2.3 Loop transfer system

This is an indirect coating system that uses loop fabric to transfer the chemicals, as shown in Fig. 8.7. The loop fabric is mounted between the immersion and bottom rollers, and loop fabric takes the liquor and transfers it to the fabrics between the two or three nip rollers (Ashish Kumar, 2009; Schindler and Hauser, 2004).

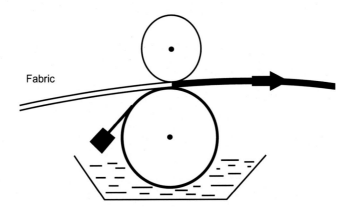

Fig. 8.5 Engraved roller.

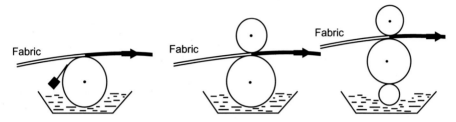

Fig. 8.6 Different kiss roll arrangements.

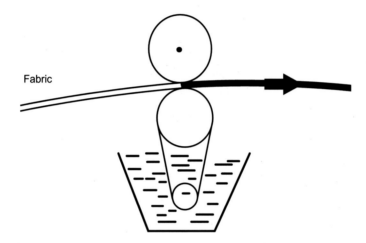

Fig. 8.7 Loop transfer coating.

8.2.2.4 Spraying system

A spray system (Fig. 8.8) is another important method among the low wet pickup methods. In this method, the chemicals or liquors are uniformly sprayed in specific amounts on the fabrics. In this system, a series of spray nozzles are used to apply or spray the finishing chemicals or liquors on the fabrics. The wet pickup of the fabric is controlled by the concentration of the finishing chemicals, the speed of the machine, the nozzle diameter, and the pressure (van der Walt and van Rensburg, 2009). Different manufacturers have developed different machines with their own merits and demerits.

8.3 Foam finishing

Foam technology is the one that consumes less water and at the same time requires less energy to dry fabrics. It drastically reduces the water consumption to 30%–90% when compared with conventional methods (Tong et al., 2012) and the method doesn't require exhaustive agents, such as common salt, in dye applications. While conventional methods of applying the finishing agents on the padding mangle attain a

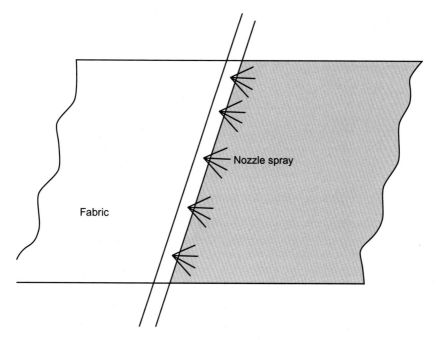

Fig. 8.8 Fabric spray coating machine.

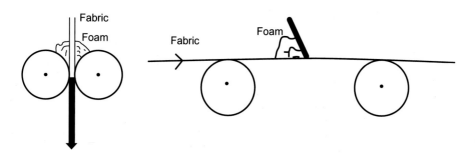

Fig. 8.9 Foam applicator.

maximum wet pickup of 60%–80%, foam finishing reduces that to the range of 25%–35% for cellulosic materials (Tong et al., 2012). The blow ratio is an important factor that decides the dilution of the finishing agents. Another important characteristic of foam that is needed for textile applications is the size of the foam. In textile applications, the smaller the size of the foam, the better the covering of the finishing agents. The most important parameter of the foam is the stability of the foam that is necessary for continuous applications. Knife over roll, kiss roller coating, vertical padding, and engraved roller coating methods are used to apply the foam along with the active finishing or dyeing agents. Fig. 8.9 illustrates the basic construction of the foam applicator (Namboodri and Duke, 1979; Bajaj, 2001).

8.4 Cold pad batch method

8.4.1 Cold pad batch method for pretreatment process

The cold pad batch (CPB) method is one of the best methods to conserve water in the textile industry and considerably reduces the processing time. In this method, the chemicals are padded in a padding mangle and then it is covered in a polyethylene paper (to avoid localized drying) and stored for 2–4 h, especially for desizing, scouring, and bleaching. This method claims 50% water savings over conventional methods (Chen, 2005, 2009).

8.4.2 Cold pad batch dyeing method

Reactive dyes are highly preferred for cotton for their excellent fastness properties and easy application. But the problem in reactive dyeing is the low exhaustion properties due to the hydrolysis of reactive dyes with water. Various techniques and methods have been implemented to overcome these issues in reactive dyeing and an effective method to dye the cotton with reactive dyes is the CPB method. This method consumes only a small amount of water and energy. The common salt used to exhaust the reactive dyes is totally eliminated in this method and it offers excellent dye fixation (Christie, 2007; Broadbent and Kong, 1995; Nitayaphat and Morakotjinda, 2017; Lewis and Seltzer, 1968; Lewis, 2009). The main disadvantage of this method is longer batching time, approximately 12 h (Broadbent and Kong, 1995).

8.5 Dyeing machines

8.5.1 Soft flow and air flow dyeing machines

In the past, winch machines were used to dye knitted fabrics, with a material to liquor ratio of 1:20. A soft flow dyeing machine needs only 4 L of water for every 1 kg of fabric processed. Now, a new range of dyeing machines is available with an $M{:}L$ ratio even lower than 4. In soft flow dyeing machines, the fabrics are circulated using the liquor, which needs some amount of liquor.

Air flow dyeing machines resemble soft flow dyeing machines, with the difference being that the fabrics are circulated using an air stream instead of the water or dye liquor used in the case of soft flow dyeing, thereby reducing the water consumption. The principle of the air flow dyeing machine is based on the aerodynamic principle instead of hydraulics. Air technology is used to transport the fabric during dyeing, washing, and even in unloading. Multinozzle air flow dyeing machines offer higher productivity and drastically reduce the material to liquor ratio to 1:2. The major advantages of soft air flow dyeing machines include (Fongs, 2018):

- flexible method to transport fabrics using air;
- able to process fabric weights ranging between 50 and 800 g per square meter;
- extremely low material to liquor ratio of 1:2 for MMF and 1:3–1:4 for natural fibers;

- nearly 40% of energy saved when compared with soft flow dyeing machines;
- reduced process time up to 25%.

8.6 Reduction of water in washing treatment

In the textile wet processing industry, a huge amount of water is used to wash the fabrics rather than in dyeing or printing. So, a reduction of water in the washing treatment saves a huge amount of water. Various techniques are used to reduce the consumption of water in the textile industry. One of these is the counter current washing technique, which is very much familiar in textile processing.

8.6.1 Counter current washing techniques

Counter current washing is one method to reduce the consumption of water a great amount. In this counter current washing method, the fabric movement is in the forward direction, whereas the water flows in the opposite direction (Fig. 8.10). This technique drastically reduces the water consumption with few modifications of the existing system. In this method, water is transferred from one trough to another by means of gravitational force. So, the second troughs are placed slightly above the first trough and the third is just above the second (Shaikh, 2009; Kocabas et al., 2009; Chougule and Sonaje, 2012).

8.7 Solvent dyeing

In recent times, there has been a focus on process sustainability and products that focus on a green philosophy, that eliminate or reduce the use or generation of hazardous compounds, and that use a minimum of energy in the process sequence. Textile wet processing leans on water and volatile organic compounds, such as solvents,

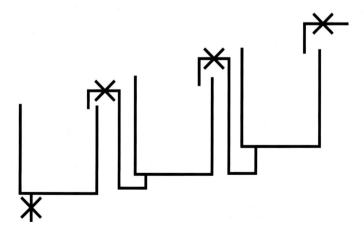

Fig. 8.10 Counter current washing technique.

surfactants, coloring matter, and oxidizing or reducing agents, which are hazardous in nature and generate a high level of pollution at disposal (Anastas and Warner, 2000). In the path toward sustainable textile manufacturing, water has been replaced with solvents to dissolve dyes to enable uniform adsorption and exhaustion onto the textile materials. Various solvents serve this purpose, indeed supercritical carbon dioxide serves better for different textile fibers with disperse dyes as the medium for the transference of dyes in and out of fiber polymer (Ahmed and El-Shishtawy, 2010); this method was initially proposed by Schollmeyer (Hasanbeigi and Price, 2015).

8.7.1 Supercritical carbon dioxide (SC-CO$_2$)

A fluid is said to be supercritical when it exhibits properties between a liquid state and a gas state (Fig. 8.11), while subjected to temperature and pressure higher than its critical point. Carbon dioxide is in a gaseous state at standard temperature (27°C) and pressure (1 atm) and it becomes solid while at low temperature. By increasing the pressure, carbon dioxide can achieve its supercritical state even at 31.1°C and 7.38 MPa (Weibel and Ober, 2002).

The high density and low viscosity of supercritical carbon dioxide lead to better solubilization and high diffusivity, respectively, and make application of SC-CO$_2$ a suitable solvent medium for the dyeing of fibers (Weibel and Ober, 2002; Saus et al., 1993) (Table 8.1).

8.7.1.1 Disperse dyes

The solubility of disperse dyes in water is moderate, as they are hydrophobic in nature, whereas in SC-CO$_2$ they exhibit low solubility, which can be increased by adding cosolvents, such as methanol and acetone, that enhance H-bonding of dyes with cosolvents (Table 8.2).

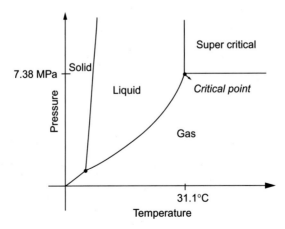

Fig. 8.11 Supercritical carbon dioxide.

Water conservation in textile wet processing

Table 8.1 Flow properties and reactivity of water, gas, and SC-CO_2 (Reid et al., 1987)

Medium	Density (g/cc)	Viscosity (g/cm/s)	Diffusion coefficient	Reactivity
Water	1	0.01	0.00005	Acts as catalyst and medium
Gas	–	0.0001	0.1	Some gases are highly reactive (CO, VOCs, NOx, NOy, SO_2, etc.)
SC-CO_2	0.6	0.0001	0.001	Inert compared to other solvents

Table 8.2 Solubility of dyes in SC-CO_2 (Guzel and Akgerman, 1999; Ferus-comelo, 2015; Laintz et al., 1991; Chang et al., 1996; Muthukumaran et al., 1999)

Disperse dyes	Solubility (mole fraction)
C. I. Mordant Brown	$1.2-5.1 \times 10^{-4}$
C. I. Mordant Yellow 12	$1.7-5.1 \times 10^{-7}$
Mordant Red 11	10^{-6}

Even though acetone is a good solvent for most chemicals, it is avoided as residual content in dyestuff as it may degrade polyester fibers at the time of dyeing. Another disadvantage is that the addition of a cosolvent may increase the cost of the process in safety measures to avoid explosions and it also reduces the reutilization of supercritical carbon dioxide as it is difficult to separate it from the cosolvents (Lin et al., 2001; Bae and Her, 1996; De Giorgi et al., 2000).

Polyester fibers dyed with disperse dye in SC-CO_2 show high dye uptake at high pressure in the range of 17–25 MPa. As the pressure increases, polymer molecules reach glass transitions at temperatures above 80°C and maximum at 120°C. In a pilot model, an airtight dyeing chamber (Fig. 8.12) is connected with a pump and expansion valve for loading carbon dioxide that can be used for dyeing polyester fabric with disperse dyes. Fastness properties also show excellent grades against washing, light, and perspiration.

8.7.1.2 Dyeing of natural fibers and blends using SC-CO_2

The supercritical CO_2 medium has attracted attention for the dyeing the cotton materials after the successful dyeing of polyester fibers with disperse dyes. Disperse dyes are suitable for dyeing cotton fibers in a supercritical carbon dioxide solvent medium because of the polar groups present in conventional dyes (direct, reactive, and vat) that exhibit poor solubility in the hydrophobic solvent SC-CO_2. Because cotton is hydrophilic, dyeing it with disperse dyes requires modification of the

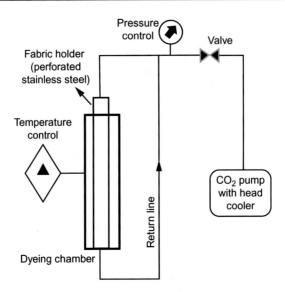

Fig. 8.12 Supercritical carbon dioxide dyeing.

cotton to have a hydrophobic nature by binding the aryl residues covalently to the cellulose molecules in the cotton fibers, which swells the fibers (Gebert et al., 1994).

$$\text{Cellulose + Sodium hydroxide} \longrightarrow \text{Cell } O^-Na^+ + C_6H_5 \text{ COCl} \longrightarrow \text{Cell} - COO - C_6H_5$$

Dyeing of modified cotton may be achieved with disperse dyes loaded with sand to distribute the bubbles and with a temperature and pressure of 100°C and 30 MPa, respectively, and dyed for 1 h with a $M{:}L$ ratio of 1:50. Modified cotton dyed with disperse dyes shows better color yield than unmodified and polyester fibers (Özcan et al., 1998). Cotton treated with the plasticizing agent polyethylene glycol also shows good fastness properties (Beltrame et al., 1998). Another novel method of dyeing cotton fibers uses reactive disperse dyes (modified disperse dye) and triethylene diamine as a catalyst (Long et al., 2012). Incorporating reactive groups, such as vinylsulfone, fluorotriazine or dichlorotriazine, 1,3,5-trichloro-2,4,6-triazine, epichlorohydrinde, and α-bromocrylic acid, into disperse dyes improves the dye uptake (van der Kraan et al., 2007; Schmidt et al., 2003; Fernandez Cid et al., 2007). But, the dichlorotriazine group damages fibers significantly during dyeing at high temperatures, while fluorotriazine reactive groups cause less damage to cotton fibers. Fernandez reported that imparting reactive groups into disperse dyes and modified hydrophobic cotton fibers yields more dye uptake and fixation percentage (85%) (Fernandez Cid et al., 2007).

Cotton and natural animal fibers, such as wool and silk, have also been experimentally dyed with modified disperse dyes having 2-reactive sites, bromoacrylic acid and 1,3,5-trichloro-2,4,6-triazine. This resulted in darker shades on wool and silk fabrics with better color fastness properties. The reverse-micellar system obtained by having the nonpolar end of the surfactant organized toward the organic solvent, and the polar

Table 8.3 Comparison of dyeing using water and SC-CO_2 medium

S. no.	Dyeing in water	Dyeing in SC-CO_2
1.	Water is main source of dyeing	Supercritical carbon dioxide fluid is used as solvent medium
2.	Huge water consumption for both dyeing and washing	Less water consumption
3.	Needs reduction treatment after dyeing	No reduction treatment after dyeing
4.	Moderate dye transfer efficiency	Efficient dye transfer
5.	Drying process consumes more energy	Material is dry and minimum energy required
6.	More effluents	Reduced effluents
7.	Longer process time, almost 3–4 h	Short duration
8.	Use of hydrose or some reducing agents causes gas emissions	Very low gas emissions

groups toward water droplets, is used in supercritical carbon dioxide dyeing of protein fibers (silk and wool) with acid dyes using ammonium carboxylate perfluoropolyether as surfactant. In another method, pentaethylene glycol n-octyl ether or 1-pentanol was used as a cosolvent to solubilize acid dyes inside of the reverse micelle declaring high low bath ratio dyeing process (Sawada and Ueda, 2004).

Dyeing polyester/cotton-fiber blend with reactive disperse dyes in a single bath reduces enormously the consumption of water and energy compared with dyeing of fiber blends (P/C blend) done at two sequential processes with two different types of dyes, dye bath solution, temperature, and time. This single step and single bath dyeing of P/C blend requires a cosolvent, such as methanol, to increase solubility and diffusion of disperse dyes in SC-CO_2 and sometimes pretreatment of both fiber blends with N-methyl-2-pyrrolidinone also improves dye uptake, but less significantly (Maeda et al., 2004) (Table 8.3).

8.7.1.3 Finishing using supercritical carbon dioxide

Biopolishing
The surface fibers of cotton fabric can be hydrolyzed by cellulase enzymes using supercritical carbon dioxide stabilized at 16.2 MPa pressure and 50°C temperature, with 48 h treatment time giving the best result (Muratov and Kim, 2002). The stereo regularity and chemical selectivity makes this supercritical carbon dioxide a unique catalyst for enzymatic finishing and processing and provides a more stable environment for enzymes in a dehydrated state (Budisa and Schulze-Makuch, 2014).

Silicone crosslinking
Cotton fabrics are treated with dimethyl siloxane polymers using supercritical carbon dioxide to impart a soft handle and silicone-based cross-linking agents (3-iso cyanate propyl tri-ethoxy silane (IPTS), tetra-ethyl ortho silicate) are used for binding silicone and fibers (Mohamed et al., 2013).

8.8 Environmental effects and waste minimization

Carbon dioxide is nonreactive, nonflammable, renewable, inexpensive, has suitable flow properties between gas and liquid, it can be 100% recovered (Hyatt, 1984), and it is an unrestricted substance by the US Environmental Protection Agency (EPA) (Anastas and Warner, 2000). The airborne concentration of carbon dioxide has a TLV (threshold limit value) at 24.85°C of 5000 ppm, which is less toxic than solvents like acetone (TLV of 750 ppm), chloroform (10 ppm) (Rearick et al., 1995; Lide, 1996). Other advantages of this fluid system are that CO_2 cannot be oxidized, it is a Lewis acid, and immune to free radicals (Beckman, 2004).

Implementation of supercritical carbon dioxide-based treatments decreases the consumption of dyes by increasing dye uptake and fixation and reduces the use of auxiliaries and chemicals. Apart from this, water use and waste water generation are reduced to a large extent along with low water pollution. This leads to smaller water treatment plants and overall energy conservation (Petek and Glavic, 1996).

Two levels of water conservation from the perspective of effluent discharge from textile wet processing are (Van der Bruggen and Braeken, 2006):

- Optimizing water consumption.
- Zero discharge or virtual zero discharge of water (and other materials), and a minimal consumption of energy.

It has been proposed to treat the wastewater from various stages of the textile chemical processing industry up to government discharge standards or to reclaim water and salt for reuse in textile wet processing. Optimizing water consumption includes the reuse of water and salt and, to some extent, other chemical auxiliaries along with the dye residual by combining the various processes or by reusing the wastewater from a process as input for another process. Others include combining the processes of scouring and bleaching, utilizing mercerization liquor to prepare sodium hypo chloride for the bleaching process, and to some extent, for cooling processes, tank cleaning, and truck cleaning (Lu et al., 2010).

The textile industry is a major consumer of water as a solvent and carrier medium for various wet processes, namely desizing, scouring, bleaching, mercerization, dyeing, printing, finishing, and washing, at all stages and leaving effluent water containing organic solvents, different dyestuffs, chemical auxiliaries, such as detergents, wetting agents, surfactants, metallic compounds, and salts. These contaminants, on discharge, increase the salinity, turbidity, BOD, COD, color intensity, and pH of waterbodies and affect the ecological system, which results in clean water starvation (Lopez et al., 1999). Zero discharge dyeing balances the environmental pollution and water starvation with the economic burdens on the textile wet processing industries by completely reusing the various processed liquors (clean water or colored water), salt, sludge, etc., with or without treating, instead of discharging them after treatment into waterbodies (Ergas et al., 2006).

For zero discharge dyeing, in the case of a batch process, the remaining dyed water after dyeing (especially with direct, reactive dyes) should be used in a jigger or winch for the next batch of process by adding dye solution, salt, and other chemical auxiliaries with subtraction of the residual concentration of the previous processed liquor. This is applicable for darker shades of colors and black in which the variation may be negligible. Implementation of this method looks to be cost beneficial and it reduces the requirement for fresh water and leads to energy saving and reduced use of chemicals. This also benefits the environmental ecosystem with lower pollutant discharges (Rajaram and Das, 2008).

The process of reclaiming the water from the collected dye liquor, decolorizing, and getting rid of the dye molecules (Table 8.4) from water are critical processes because the structure differs for different types of dyes but falls under either anionic, cationic, or nonionic categories.

Table 8.4 Decolorizing methods for dyes in waste water (Lopez et al., 1999; Ergas et al., 2006; Robinson et al., 2001; Tang and Chen, 2002)

Principle	Methods	Description
Chemical oxidation	H_2O_2 at UV light	Aromatic ring cleavage gives colorless compounds, and reduces sludge and foul odors
	Fentons reagent $H_2O_2 - Fe(II)$	Removal of dyes by sorption or bonding
	Ozonation	Instability of O_3 breaks the dyes into smaller ones. No sludge formation and low COD of effluent on discharge
	Sodium hypochloride	Cl^- attacks the amino groups in dyes, causes break of azo bond
Physical adsorption	Activated carbon	Suitable adsorbent for cationic, mordant, and acid dyes
	Peat	Adsorbs polar organic compounds, doesn't need activation process
	Fly ash	Gives more surface area and mostly combined with activated carbon
	Silica gel	Especially used for basic dyes, commercially unsuccessful
	Wood chips	Good adsorbent for acid dyes, but ineffective as more contact time is required
	Membrane filtration	Separates dye particles and suitable for low concentrations
	Ion exchanger	Separates both water soluble and insoluble dyes from effluents by ion exchange until saturation
Biological decolorization	White rot fungi	*P. chrysosporium* decolorize the dyes using lignin peroxidase and manganese peroxidase enzymes
	Anaerobic dye bioremediation systems	Azo and other water-soluble dyes are decolorized

The Fenton's agent and ozonation oxidize the color even if the effluent is highly concentrated. Reverse osmosis is a process in which particles are filtered through osmotic pressure and nanofiltration is used to recover the water; it is successful for around 87%–90% recovery depending on the amount of effluent and concentration and the water is reused in the dyeing process (Asano and Levine, 1996). Screening, primary physico-chemical treatments followed by ozonation, and biological treatments (reduces COD) are the pretreatment processes that make the effluent fit for reverse osmosis, avoiding membrane fouling and improving maintenance, performance, and life (Rearick et al., 1995; Nataraj et al., 2009).

Though these methods offer high value propositions and economic benefits, problems include awareness on the part of the processors and the ability to replace the existing processing sequence machinery with these machines that need high investment, particularly in the initial costs. It calls for a conscious effort by the manufacturers of both machinery as well as fabrics to arrive at an amicable solution for systematic replacement of processes that involve a high water footprint.

8.9 Conclusion

The textile processing industry needs water to process textile goods. Due to the scarcity of water, new methods and processes have been developed to reduce or eliminate the use of water. By using these techniques and methods, such as low wet pickup techniques, foam processing, solvent dyeing and finishing, the CPB method, the latest air flow dyeing machines, counter current washing methods, and the zero discharge method, the consumption of water is reduced and this paves the way for sustainable manufacturing practices.

References

Ahmed, N.S.E., El-Shishtawy, R.M., 2010. The use of new technologies in coloration of textile fibers. J. Mater. Sci. 45, 1143–1153. https://doi.org/10.1007/s10853-009-4111-6.

Anastas, P.T., Warner, J.C., 2000. Green Chemistry: Theory and Practice. Oxford University Press, New York.

Asano, T., Levine, A.D., 1996. Wastewater reclamation, recycling and reuse: past, present, and future. Water Sci. Technol. 33, 1–14.

Ashish Kumar, S., 2009. Coated Textiles: Principles and Applications. CRC Press.

Bae, H.K., Her, B.K., 1996. Solubility of disperse dyes in supercritical carbon dioxide. Hwahak Konghak. .

Bajaj, P., 2001. Finishing of textile materials. J. Appl. Polym. Sci. 83, 631–659. https://doi.org/10.1002/app.2262.

Beckman, E.J., 2004. Supercritical and near-critical CO_2 in green chemical synthesis and processing. J. Supercrit. Fluids 28, 121–191. https://doi.org/10.1016/S0896-8446(03)00029-9.

Beltrame, P.L., Castelli, A., Selli, E., et al., 1998. Dyeing of cotton in supercritical carbon dioxide. Dyes Pigments 39, 335–340. https://doi.org/10.1016/S0143-7208(98)00016-3.

Broadbent, A.D., Kong, X., 1995. The application of reactive dyes to cotton by a wet-on-wet cold pad-batch method. J. Soc. Dye Colour. 111, 187–190. https://doi.org/10.1111/j.1478-4408.1995.tb01721.x.

Broadbent, B., Quebec, S., 1990. Basic principles of vacuum slot extraction. AATCC Rev. 22, 13–17.

Budisa, N., Schulze-Makuch, D., 2014. Supercritical carbon dioxide and its potential as a life-sustaining solvent in a planetary environment. Life 4, 331–340. https://doi.org/10.3390/life4030331.

Chang, K.-H., Bae, H.-K., Shim, J.-J., 1996. Dyeing of pet textile fibers and films in supercritical carbon dioxide. Korean J. Chem. Eng. 13, 310–316. https://doi.org/10.1007/BF02705955.

Chen, L., 2009. Cold pad-batch technique of energy conservation and emission reduction. Text. Finish. J. 31, 51–54.

Chen, M., 2005. Study on Water-Conservation Measures Cleaner Production of Printing and Dyeing Industry in Zhejiang. Zhengjiang University, Hangzhou.

Chougule, M.B., Sonaje, N.P., 2012. Novel techniques of water recycling in textile wet processing through best management practices (BMP's). Int. J. Appl. Sci. Adv. Technol. 1, 29–33.

Christie, R.M., 2007. Environmental Aspects of Textile Dyeing. Woodhead Publishing Limited.

De Giorgi, M.R., Cadoni, E., Maricca, D., Piras, A., 2000. Dyeing polyester fibres with disperse dyes in supercritical CO_2. Dyes Pigments 45, 75–79. https://doi.org/10.1016/S0143-7208(00)00011-5.

Ergas, S.J., Asce, M., Therriault, B.M., et al., 2006. Evaluation of water reuse technologies for the textile industry. J. Environ. Eng. 132, 315–323.

Fernandez Cid, M.V., van Spronsen, J., van der Kraan, M., et al., 2007. A significant approach to dye cotton in supercritical carbon dioxide with fluorotriazine reactive dyes. J. Supercrit. Fluids 40, 477–484. https://doi.org/10.1016/j.supflu.2006.07.011.

Ferus-comelo, M., 2015. A new method to measure the solubility of disperse dyes in water at high temperature. Color. Technol. 269–271. https://doi.org/10.1111/cote.12153.

Fongs, 2018. Then-airflow-synergy. In: Fongs.http:/www.fongs.eu/12/then/then-airflow-synergy.html. Accessed 26 February 2018.

Gebert, B., Saus, W., Knittel, D., et al., 1994. Dyeing natural fibers with disperse dyes in supercritical carbon dioxide. Text. Res. J. 64, 371–374. https://doi.org/10.1177/004051759406400701.

Goldstein, H.B., Smith, H.W., 1980. Lower limits of low wet pick-up finishing. Text. Chem. Color. 12, 27.

Guzel, B., Akgerman, A., 1999. Solubility of disperse and mordant dyes in supercritical CO(2). J. Chem. Eng. Data 44, 83–85. https://doi.org/10.1021/je980157v.

Hasanbeigi, A., Price, L., 2015. A technical review of emerging technologies for energy and water efficiency and pollution reduction in the textile industry. J. Clean. Prod. 95, 30–44. https://doi.org/10.1016/j.jclepro.2015.02.079.

Hyatt, J.A., 1984. Liquid and supercritical carbon dioxide as organic solvents. J. Org. Chem. 49, 5097–5101. https://doi.org/10.1021/jo00200a016.

Kocabas, A.M., Yukseler, H., Dilek, F.B., Yetis, U., 2009. Adoption of European Union's IPPC Directive to a textile mill: analysis of water and energy consumption. J. Environ. Manag. 91, 102–113. https://doi.org/10.1016/j.jenvman.2009.07.012.

Laintz, K.E., Wai, C.M., Yonker, C.R., Smith, R.D., 1991. Solubility of fluorinated metal diethyldithiocarbamates in supercritical carbon dioxide. J. Supercrit. Fluids 4, 194–198. https://doi.org/10.1016/0896-8446(91)90008-T.

Leah, R.D., 1982. Low wet pick-up processing for the 80s. JSDC 98, 422–429.

Lewis, D.M., 2009. Letter to the Editor Research in the cold pad-batch dyeing process for wool pretreated by hydrogen peroxide. Color. Technol. 125, 304. https://doi.org/10.1111/j.1478-4408.2009.00211.x.

Lewis, D.M., Seltzer, I., 1968. Corn mu n ications pad-batch dyeing of wool with reactive dyes. Color. Technol. 84, 501–507.

Lide, D.R., 1996. Handbook of Chemistry and Physics, Seventy-Si. CRC, Boca Raton, FL.

Lin, H.M., Liu, C.Y., Cheng, C.H., et al., 2001. Solubilities of disperse dyes of blue 79, red 153, and yellow 119 in supercritical carbon dioxide. J. Supercrit. Fluids 21, 1–9. https://doi.org/10.1016/S0896-8446(01)00075-4.

Long, J.J., Xiao, G.D., Xu, H.M., et al., 2012. Dyeing of cotton fabric with a reactive disperse dye in supercritical carbon dioxide. J. Supercrit. Fluids 69, 13–20. https://doi.org/10.1016/j.supflu.2012.05.002.

Lopez, A., Ricco, G., Ciannarella, R., et al., 1999. Textile wastewater reuse: ozonation of membrane concentrated secondary effluent. Water Sci. Technol. 40, 99–105.

Lu, X., Liu, L., Liu, R., Chen, J., 2010. Textile wastewater reuse as an alternative water source for dyeing and finishing processes: a case study. Desalination 258, 229–232. https://doi.org/10.1016/j.desal.2010.04.002.

Maeda, S., Kunitou, K., Hihara, T., Mishima, K., 2004. One-bath dyeing of polyester/cotton blends with reactive disperse dyes in supercritical carbon dioxide. Text. Res. J. 74, 989–994. https://doi.org/10.1177/004051750407401109.

Mohamed, A.L., Er-Rafik, M., Moller, M., 2013. Supercritical carbon dioxide assisted silicon based finishing of cellulosic fabric: a novel approach. Carbohydr. Polym. 98, 1095–1107. https://doi.org/10.1016/j.carbpol.2013.06.027.

Muratov, G., Kim, C., 2002. Enzymatic hydrolysis of cotton fibers in supercritical CO_2. Biotechnol. Bioprocess. Eng. 7, 85–88. https://doi.org/10.1007/BF02935884.

Muthukumaran, P., Gupta, R.B., Sung, H.-D., et al., 1999. Dye solubility in supercritical carbon dioxide. Effect of hydrogen bonding with cosolvents. Korean J. Chem. Eng. 16, 111–117. https://doi.org/10.1007/BF02699013.

Namboodri, C.G., Duke, M.W., 1979. Foam finishing of cotton-containing textiles. Text. Res. J. 49, 156–162. https://doi.org/10.1177/004051757904900308.

Nataraj, S.K., Hosamani, K.M., Aminabhavi, T.M., 2009. Nanofiltration and reverse osmosis thin film composite membrane module for the removal of dye and salts from the simulated mixtures. Desalination 249, 12–17. https://doi.org/10.1016/j.desal.2009.06.008.

Nitayaphat, W., Morakotjinda, P., 2017. Cold pad-batch dyeing method for cotton fabric dyeing with Uncaria gambir bark using ultrasonic energy. Chiang Mai J. Sci. 44, 1562–1569. https://doi.org/10.1016/j.ultsonch.2011.04.001.

Özcan, A.S., Clifford, A.A., Bartle, K.D., Lewis, D.M., 1998. Dyeing of cotton fibres with disperse dyes in supercritical carbon dioxide. Dyes Pigments 36, 103–110. https://doi.org/10.1016/S0143-7208(97)00005-3.

Petek, J., Glavic, P., 1996. An integral approach to waste minimization in process industries. Conserv. Recycl. 17, 169–188. https://doi.org/10.1016/0921-3449(96)01151-2.

Rajaram, T., Das, A., 2008. Water pollution by industrial effluents in India: discharge scenarios and case for participatory ecosystem specific local regulation. Futures 40, 56–69. https://doi.org/10.1016/j.futures.2007.06.002.

Rearick, W.A., Farias, L.T., et al., 1995. Water and salt reuse in the dyehouse. Text. Chem. Color. 29, 10–19.

Reid RC, Prausnitz JM, Poling BE (1987) The Properties of Gases and Liquids. United States

Robinson, T., McMullan, G., Marchant, R., Nigam, P., 2001. Remediation of dyes in textile effluent: a critical review on current treatment technologies with a proposed alternative. Bioresour. Technol. 77, 247–255. https://doi.org/10.1016/S0960-8524(00)00080-8.

Yoshikazu Sando, Hiroshi Shidoshiro (1981) Apparatus for supplying a definite amount of a treating liquid to a textile product continuously, 6.

Saus, W., Knittel, D., Schollmeyer, E., 1993. Dyeing of textiles in supercritical carbon dioxide. Text. Res. J. 63, 135–142. https://doi.org/10.1177/004051759306300302.

Sawada, K., Ueda, M., 2004. Evaluation of the dyeing mechanism of an acid dye on protein fibers in supercritical CO_2. Dyes Pigments 63, 77–81. https://doi.org/10.1016/j.dyepig.2004.01.008.

Schindler, W.D., Hauser, P.J., 2004. Chemical Finishing of Textiles. Woodhead Publishing Ltd, Cambridge.

Schmidt, A., Bach, E., Schoomeyer, E., 2003. The dyeing of natural fibres with reactive disperse dyes in supercritical carbon dioxide. Dyes Pigments 56, 27–35. https://doi.org/10.1016/S0143-7208(02)00108-0.

Shaikh, M., 2009. Water conservation in textile industry. Ptj 48, 48–51.

Tang, C., Chen, V., 2002. Nanofiltration of textile wastewater for water reuse. Desalination 143, 11–20. https://doi.org/10.1016/S0011-9164(02)00216-3.

Tong, O., Shao, S., Zhang, Y., et al., 2012. An AHP-based water-conservation and waste-reduction indicator system for cleaner production of textile-printing industry in China and technique integration. Clean Techn. Environ. Policy 14, 857–868. https://doi.org/10.1007/s10098-012-0453-x.

Van der Bruggen, B., Braeken, L., 2006. The challenge of zero discharge: from water balance to regeneration. Desalination 188, 177–183. https://doi.org/10.1016/j.desal.2005.04.115.

van der Kraan, M., Fernandez Cid, M.V., Woerlee, G.F., et al., 2007. Dyeing of natural and synthetic textiles in supercritical carbon dioxide with disperse reactive dyes. J. Supercrit. Fluids 40, 470–476. https://doi.org/10.1016/j.supflu.2006.07.019.

van der Walt, G.H.J., van Rensburg, N.J.J., 1984. A review of low add-on and foam application techniques. SA WTRI Spec. Publ. 1–59.

van der Walt, G.H.J., van Rensburg, N.J.J., 2009. Low-liquor dyeing and finishing. Text. Prog. 14, 37–41. https://doi.org/10.1080/00405168608688900.

Weibel, G.L., Ober, C.K., 2002. An overview of supercritical CO_2 applications in microelectronics processing. Microelectron. Eng. 65, 145–152. https://doi.org/10.1016/S0167-9317(02)00747-5.

Water requirement and sustainability of textile processing industries

A.S.M. Raja, A. Arputharaj, Sujata Saxena, P.G. Patil
Chemical and Biochemical Processing Division, ICAR-Central Institute for Research on Cotton Technology, Mumbai, India

9.1 Introduction

Globally, the textile industry plays a major role in the economic, social, and cultural development of nations. It is concerned with every human being in the world. The different sectors of the textile industry include fiber production, spinning, weaving, dyeing and finishing, garmenting, and denim, among others. Apart from conventional processing industries, nowadays new areas, such as nonwoven, composites, technical textiles, medical textiles, and industrial textiles, are also emerging rapidly. It is estimated that more than 60 million people are directly employed in the textile industry (Anon, 2015). In the global textile industry, the countries that play a predominant role include China, United States, India, Pakistan, Brazil, Indonesia, Taiwan, Turkey, Bangladesh, and South Korea. Many processes and products that go into the making of fibers, textiles, apparel, and technical textile products consume significant quantities of water and fossil fuel. Apparel and textiles account for approximately 10% of the total global carbon impact (Conca, 2015). The estimated consumption of electricity and water for an annual global production of 60 billion kilograms of fabrics is approximately 1 trillion kilowatt hours and 9 trillion liters of water (Zaffalon, 2010). Hence, textile-based industries are prime targets of the environmentalists in their crusade against pollution.

This chapter briefly discusses multiple aspects of water requirements and different approaches to reduce and recycle water use in the textile industry. The importance of making conventional textile processing sustainable and ecofriendly is discussed in detail. The amounts of water used in cotton and the textile industry and the water use at each processing step are concisely highlighted to show the need for developing water-conserving and waterless technologies. Water-conserving techniques and effective machinery can be used to reduce water wastage. This chapter also gives an exhaustive overview of all the waterless systems that have been developed, such as supercritical CO_2-based processing, plasma processing, zero liquid discharge (ZLD) system, and digital printing. All of the aspects of using water sustainably, conservation, and waterless systems are meticulously discussed further.

9.2 Sustainability

Sustainable development is a complex multidimensional concept concerning the environment, economy, human health, and social impact. Sustainable development can be defined in different ways. Sustainability is economic development that is conducted without depletion of natural resources. The widely used definition of sustainability is, "development that meets the needs of the present without compromising the ability of future generations to meet their own needs," by the Brundtland Commission report on Our Common Future in the year 1987. In the year 1800 CE, the world population was about 1 billion. Since that time, the world population has been growing too quickly. During the 1970s, the world population was 4 billion and now it is estimated to be approximately 7.4 billion. In order to satisfy the food, clothing, shelter, and energy needs of the growing world population, sustainable use of water is very important. Development cannot be sustained because of the deterioration of the Earth's most precious resource, water.

The need for sustainability arises due to the following factors:

- Global warming
- Greenhouse effect
- Ozone layer depletion
- Climate change, unpredictable weather
- The rise in sea level
- Unpredictable rain patterns
- Natural calamities
- Disappearance of glaciers

The increase in the average temperature of the earth is the natural phenomenon. The warming of earth facilitated the creation of life. However, after the industrial revolution of the 18th century, the rate at which the average earth temperature rose was too fast and this is known as global warming. This artificial global warming is due to the release of greenhouse gases, such as carbon dioxide and chlorofluorocarbon and is due to the use of fossil fuels and other developments. The average Earth temperature has risen by 0.6–0.8°C according to different estimates. The rise in temperature is approximately 10 times faster than the normal rise. Global warming leads to ozone layer depletion, climatic change, and unpredictable weather. Ozone layer depletion refers to the removal of the natural ozone layer that is present above 15–20 km away from the Earth, mainly due to the emission of chlorofluorocarbon gases. Depletion of the ozone layer leads to harmful radiation being incident on the Earth, which results in the extinction of many living organisms. The dangerous fallout of global warming is climate change. As climate change warms the atmosphere, it leads to an increase in the demand for water, uneven rain patterns, salt water intrusion, and water quality impairment. All the above-mentioned factors ultimately affect the water available for human beings.

9.2.1 Sustainable designs for water

The water available in the earth is continuously decreasing and the demand for the same is increasing due to increasing population, human needs, and climate change. If the situation is allowed to continue, very soon a tipping point will be reached at which a severe water crisis and associated human conflicts over water could happen.

In order to avert this, sustainable design and development have to be attempted. Sustainable design should address the following points:

1. Decreasing the demand for water by use of better science practices.
2. Replenishing the water for reuse.

9.3 Water availability in India

India has 18% of the world population but it has only 4% of the world's water resources. Average annual water availability in India is 1869 BCM (billion cubic meter), but the average potential of the useable quantity of water is 1123 BCM, which consists of 690 BCM surface water and 433 BCM groundwater. Annual per capita availability of water is 1545 cubic meter compared to 1818 in 2001 and 6042 in 1947. The figures indicate that the availability of water is steadily decreasing. The latest assessment of the country's dynamic groundwater resources performed by the Central Groundwater Board (CGWB) showed that consumption of groundwater is much higher than their rechargeable limit every year, which will lead to severe water scarcity in the near future. In this context, the water used by the textile industry in relation to methods to conserve the water is discussed in this article (Water and Related Statistics, 2010).

India is one of the largest producers of textiles and garments. The major growing textile sector contributes 5% of GDP and 14% of the industrial output index and 13% of the country's export earnings. It gives direct employment to 45 million people, second only to agriculture. It is estimated that the current market size of the textile industry in India is 120 billion USD comprising 80 billion USD of domestic consumption and 40 billion USD of export. The global share of the Indian textile industry is 5% and it is growing at a CAGR of 7%. Cotton fiber is the backbone of the Indian textile industry with a share of 60% (Confederation of Indian Industries, 2018).

The different segments of the Indian textile industry are listed below.

1. Cotton segment
2. Organized mills
3. Manmade fiber and filament yarn industry
4. Wool and woolen textiles
5. Sericulture and silk textiles
6. Jute and jute textiles
7. Powerlooms
8. Handlooms
9. Handicrafts
10. Apparel and garments

The Indian textile industry consumes a diverse range of fibers with cotton as the predominant fiber. The ratio between the cotton and manmade fiber/filament is calculated as 59:41. More than 3400 textile mills are present in the organized textile sector. India has the highest number of spindles (about 50 million) and 850,000 rotors. Functionally, the Indian textile industry can be classified into the:

1. Spinning industry
2. Weaving industry

3. Processing industry
4. Composite mills
5. Garment and garment processing industries

In India, the textile processing industry is a leading consumer of water and it ranks among the top ten water-consuming industries (Agarwal and Kumar, 2011). The leading water consuming industries are given in Fig. 9.1.

Although the water consumed by the textile industry is comparatively low, the pollution created by the industry can contaminate large amounts of water resources and agriculture.

9.3.1 Water requirement for textile processing industry

An average composite textile mill, which produces 8000 kg/day of fabric, consumes roughly 1.6 million liters of water per day, out of which 8% is utilized in printing and 16% is used in the dyeing process (Kant, 2012). Mechanical processing of textiles, such as spinning, weaving, and garmenting, requires very little water. However, the chemical processing operations, such as scouring, bleaching, dyeing, and finishing, require a substantial quantity of water. After processing, the used water is released as effluent; if it is untreated, the effluent can contaminate groundwater resources. In textile wet processing, water is used mainly for three purposes, namely, as a solvent for dyes and chemicals, as a medium for transferring dyes and chemicals to fabric, and as a washing and rinsing medium. Apart from the above processes, ion exchange, boiler, cooling water, steam drying, and the cleaning part of the process also consume a considerable amount of water.

There are three key areas of water use in textile processing

- Fabric or yarn precleaning and rinsing prior to dyeing
- Dyeing or printing operation and rinsing

Fig. 9.1 Water consumption in different industries (Agarwal and Kumar, 2011).

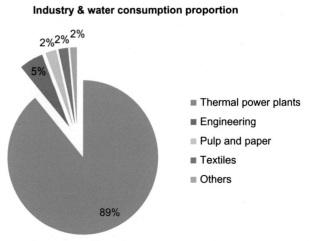

- Finishing
- Garment washing

The amount of water used varies widely depending upon the type of textile fiber processed, the type of product, for example, woven or knitted, and the specific processes and equipment. The quantity of water used for processing varies depending upon the type of fiber. Generally, natural fibers, such as cotton and wool, require more water for processing than synthetic fibers, such as polyester and nylon. The water requirement for processing 1 kg of cotton or polyester fabric is given in Table 9.1. Natural fibers are hydrophilic in nature and, generally, water is used as dyeing medium for the exhaustion of dye on the substrate. During processing, water is absorbed by these fibers along with dyes and chemicals. Therefore, more water is required for processing natural fibers. In contrast, synthetic fibers are generally hydrophobic and do not absorb much water. For some fibers, such as polyester, the thermal fixation technique is used to exhaust the dye inside the fiber structure.

The exact amount of water used by the textile wet processing industries is difficult to calculate because of a great diversity of products (woven, knitted, nonwoven, etc.), processes, machinery used, and of dyes and chemicals. Table 9.2 gives rough estimates of the water required for processing 1 kg of material for different fibers.

9.3.2 Water requirements for the cotton processing industries

The cotton textile processing industry can be broadly divided into four categories:

1. Woven processing
2. Knit processing
3. Garment processing
4. Nonwovens and miscellaneous processing, such as absorbent cotton and surgical gauze

Each category of industry follows different unit operations for processing. The general structure of the unit operations is given in Table 9.3.

Table 9.1 Comparative water requirement for 1 kg cotton and polyester material (Menezes and Mrinal, 2011)

Process	Consumption in L/kg	
	Cotton	Polyester
Sizing	0.5–8.0	–
Desizing	5–20	–
Scouring and bleaching/prewashing	20–50	10–20
Mercerizing	15–30	–
Dyeing	15–30	17–34
Printing	8–16	8–16
Finishing	10–20	8–12

Table 9.2 Approximate requirement of water for processing of different textile fibers (Arputharaj et al., 2015)

Type of material	Required quantity of water (L/kg)	Required quantity of water (L/kg)
	MLR[a] 1:10–15	MLR 1:5–9
Cotton	80–200	40–100
Wool	100–300	50–150
Polyester	40–100	40–80
Nylon	50–120	50–100
Acrylic	30–80	30–80

[a]MLR: material to liqour.

Table 9.3 Unit operations involved in different categories

Woven	Knitted	Garment	Nonwoven
Sizing	Prewashing/ demineralization	Prewashing/desizing	Scouring
Desizing	Scouring and bleaching	Scouring and bleaching	Washing
Washing	Washing	Washing	Bleaching
Demineralization	Dyeing	Dyeing	Washing
Washing	Washing	Washing	
Scouring	Different finishing operations	Enzyme washing/stone washing	
Washing	Washing	Different finishing application	
Peroxide bleaching			
Washing			
Dyeing			
Washing			
Different finishing operations			
Washing			

Even though the unit operations are similar, depending upon the category of materials, different types of machine and process sequences are involved. For example, typical knit processing requires more water than woven processing for the same unit preparation.

9.3.3 Process-wise water consumption

As mentioned earlier, chemical processing can be generally classified into three stages:

Table 9.4 The process-wise percentage water requirement (Arputharaj et al., 2015)

Process	% water requirement
Scouring and bleaching	38–50
Dyeing	20–30
Printing	8–10
Finishing	8–10
Humidification	5–9

1. Preparatory processes—desizing, scouring, bleaching
2. Dyeing
3. Finishing

Table 9.4 gives the process-wise water requirement. From the table, it is very clear that preparatory processes of textiles, namely, desizing, scouring, and bleaching, which require several steps of washing, are highly water-consuming operations. During preparatory processes, natural and added impurities are removed from the textile materials (Yarns and fibres.com, 2010).

9.3.4 Water requirement for Indian cotton sector

The water requirement in the cotton sector is calculated based on the production figures of cotton and blended yarn production during the year 2013–14. In that year, India produced approximately 7110 million kg of yarn, comprising cotton, blended, and filament yarns (Ministry of Textiles, 2018). Assuming that 80% of the total produced yarn was dyed, the water used for processing was calculated and is given in Table 9.5. Based on our rough estimates, approximately 515 million cubic meters of water was used for processing. Hence, there is huge scope to save water during processing.

9.4 Water-conserving techniques for textile wet processing

Significant reductions in water use can be achieved by preventing unnecessary water consumption in textile processing mills. Implementation of the best available techniques (BAT) should be employed for achieving significant reductions in water use and effluent generation during processing. It should be noted that energy consumption is directly related to water consumption as most of the energy is required to heat up the process baths. Therefore, a reduction in water consumption results in a reduction in energy consumption also. There are several new technologies available to conserve water in the textile processing industry. They can be classified into:

- Effective chemical management, which includes textile auxiliaries, nanochemicals, high-exhaust dyes, etc.

Table 9.5 Water requirement in the cotton sector

Particulars	In million kg	Water for processing 1 kg of material in cubic meters	Total water requirement in million cubic meters
Cotton Yarn production	4055		
20% white	811	0.03	24.33
80% dyed	3244	0.1	324.4
Blended cotton yarn	1433		
20% white	286.6	0.02	5.732
80% dyed	1146.4	0.07	80.248
Manmade filament yarn	1622	0.05	81.1
Total			515.81

- Development of efficient machines for processing, which require ultra-low liquor ratios.
- Waterless processing techniques, such as plasma processing, supercritical CO_2, and laser.

9.4.1 Process and chemical management

Desizing: The object of desizing is to remove size that is applied to the gray fabric as a part of the weaving process so that the fabric becomes free of the added matter and can be effectively scoured and bleached. The methods of desizing include hydrolytic (rot steeping, acid steeping, and enzymatic) and oxidative (chlorine, chlorite, bromite). Out of several methods, adoption of the enzymatic process can lead to water saving due to a lower number of wash cycles.

Scouring: This process is carried out to remove natural as well as added impurities of essentially hydrophobic character (oil, fats, waxes, etc.) as completely as possible and leave the fabric in a highly absorptive condition without undergoing significant chemical or physical damage.

Bleaching: In this process, the natural coloring matter and any other coloring matter is removed from the natural fiber. However, for light and medium shades or when fiber brightness is important, bleaching becomes an essential operation before dyeing or printing. The process can be carried out using two different bleaching agents, such as oxidative (chlorine, hydrogen peroxide, hypochlorous acid, etc.) or reductive (sulfur dioxide, sodium sulfite, sodium hydrosulfite).

Conventionally, scouring and bleaching were done as separate processes. After the introduction of the hydrogen peroxide bleaching process, which also requires an alkaline condition, most of the industry now adopts a combined scouring and bleaching process. As mentioned earlier, the preparatory process consisting of desizing, scouring, and bleaching consumes a significantly higher amount of water. Because the

Table 9.6 Percentage of unfixed dyes in effluent (GG 62, 1997)

Class of dye	Percentage of unfixed dye
Azoic dyes	5–10
Reactive dyes	20–50
Direct Dyes	5–20
Vat dyes	5–20
Pigment	1–2
Sulfur dyes	30–40

commonly employed alkaline hydrogen peroxide-based pretreatment requires a large amount of water, research has to be initiated to explore the possibility of alternate processes. The use of plasma for removing impurities, such as size, wax, and residual color, is a promising alternative to completely eliminate the use of water in the pretreatment process.

Dyeing: Dyeing is the process of adding color to textile products, such as fibers, yarns, and fabrics. Dyeing is normally done in a special solution containing dyes and particular chemical materials. Cotton material can be dyed with different classes of dyes, such as reactive, direct, vat, and sulfur dyes. Every dye has a different exhaustion rate on the textile materials. The percentages of unfixed dyes present in the effluent for different classes of dye are given in Table 9.6.

The unfixed dyes have to be removed by employing a number of washing cycles. Reactive dyes, which are predominately used for dyeing cotton, exhibit only 70% exhaustion. The reactive dyed cotton materials have to be soaped and washed several times, which consumes a huge quantity of water. Unfixed dyes also increase the pollution load and effluent treatment. Hence, in order to reduce the water requirement, efforts have to be made to increase the exhaustion of dye on the materials. If we are able to increase the exhaustion to more than 90%, the water requirement for dyeing can be reduced approximately 20%–30%.

The equipment employed also significantly influences the water usage. Table 9.7 gives information about the amount of water to be taken per gram of cotton material, popularly called the material-to-liquor ratio (MLR), for various equipment used in processing (Yarns and fibres.com, 2010).

The liquor ratio is the major factor influencing water consumption. Table 9.8 shows the influence of processing at 1:10 and 1:7 liquor ratios for knitted goods.

Therefore, by employing a machine requiring a lower MLR, a 30% reduction of water consumption can be achieved. Another point to be noted from the above table is that, out of the total water consumption, 50% of the water is required for washing purposes only. Development of effective washing techniques can save water. Combining different processes and employing chemicals that require fewer washing cycles for removal are some of the methods that can be adopted to reduce water use. Combined scouring and bleaching using alkaline hydrogen peroxide is one of the techniques successfully adopted in the industry. Developments in enzyme-based processes, which can reduce the number washing cycles, have also been adopted by the industry.

Table 9.7 Material-to-liquor ratio in different equipment (Yarns and fibres.com, 2010)

Dyeing machine	MLR
Continuous	1:1
Winch	1:15 to 1–40
Soft flow	1:5 to 1:10
Air Jet flow	1:2 to 1:5
Jet	1:7 to 1:15
Jig	1:5
Beam	1:10
Package	1:10
Beck	1:17

Table 9.8 Water consumption as affected by liquor ratio used in different processing stages

	Water consumption/liquor ratio	
Processing stage	1:10	1:7
Desizing	10	7
Washing	10	7
Scouring and bleaching	10	7
Washing	10	7
Neutralization	10	7
Washing	10	7
Reactive dyeing	10	7
Washing	10	7
Soaping	10	7
Washing	10	7
Finishing	10	7
Total requirement of water	110	77

9.4.2 Effective dyeing machinery

Effective dyeing machines that require only very low liquor for processing and that also ensure effective interaction between the material and the chemicals are the need of the hour. In the recent past, machinery used by the industry required a MLR as high as 1:30. However, due to developments in the machinery, industry is now adopting ultra-low liquor machinery (ULLR), which requires only 1:3–1:9. The knit processing segment requires more water due to the adoption of the batch-wise process. Nowadays, continuous processing ranges have been developed, which require very low water compared to the batch-wise process. Currently for the batch-wise process, ULLR dyeing equipment is also available on the market for the processing of textile material using batch methods.

Table 9.9 Amount of water required for processing 1 kg cotton and blended materials using continuous and batch-wise processes (Zietlow, 2013)

Process	Water for processing 1 kg of fabric
Polyester knitted goods, *continuous* prewashing, and *discontinuous* dyeing	80
Cotton desizing, bleaching, and dyeing using continuous process	75
Cotton cord fabric, desizing, and hot bleaching *continuously* in two steps, reactive dyeing in *discontinuous* technique (Jigger)	120
Dyeing and printing, only *continuous technique*, mainly cellulosic fabrics, reactive and vat dyeing and printing	50
Polyester/cotton dyeing, *continuous technique* with polyester followed by continuous reactive or vat dyeing on pad-steam-aggregate	40

9.4.2.1 Batch-wise machines versus continuous range

Water use can be significantly reduced by adopting continuous processing machines instead of batch-wise processes. The production rate can also be increased with less water, chemicals, energy, and labor. Table 9.9 gives the approximate amount of water required for processing 1 kg cotton and blended materials using the continuous and batch-wise processes.

9.4.3 Other water-conserving techniques

9.4.3.1 General technical measures

Good housekeeping in the mill will reduce the water use by 10%–30%. Good housekeeping measures are often carried out without significant investments, but they lead to substantial cost savings and the saving of water, chemicals, and energy. Good housekeeping measures are essential for a company that is critical about its own behaviors. Implementing the following can make significant reductions in water use, minimizing leaks and spills (Zietlow, 2013).

- Plugging leaks and checks on running taps.
- Installation of water meters or level controllers on major water-carrying lines.
- Turn off the water when machines are not operating.
- Identifying unnecessary washing of both fabric and equipment.
- Training employees on the importance of water conservation.

As the water scarcity and water pollution problems caused by industrial manufacturing are becoming increasingly serious globally, industrial manufacturers are forced to adopt water conservation techniques and pollution abatement technologies, especially in the textile wet processing industry, which is one of the highly water intensive industries in the world. In order to support textile manufacturers, it is very important to develop water conservation and pollution controls and cleaner production technologies more systematically and comprehensively (Chen et al., 2016).

Ozturk et al. (2016) reported that after the implementation of the following BAT in Turkey, the cotton/polyester fabric dyeing industry achieved reductions of 43%–51% in water consumption, 16%–39% in chemical consumption, 45%–52% in combined wastewater flowrate, and 26%–48% in specific chemical oxygen demand load.

- Reuse/recovery of washing/rinsing and softening wastewater
- Reuse of suitable dye bath
- Optimization of water softening unit
- Recovery/reuse of regeneration wastewater
- Application of counter-washing techniques in the pad-batch washing process

Guyer et al. (2016) reported that countercurrent washing-based strategies and advanced oxidation process (AOP) treatment of textile effluent can save a good amount of wastewater in the textile industry. They studied the recyclability and direct reuse of washing/bleaching wastewater from reactive dyeing of cotton fabric through (AOPs) (O_3, UV-O_3, O_3-H_2O_2, and O_3-H_2O_2-UV) in the Turkish textile industry. Dyeing trials with treated wastewater were carried out to find the reuse potential of the AOP-treated wastewater without having any adverse effect on fabric quality. They found that 95%–100% color removal was possible through applied AOPs for 30 min detention time, while COD removal through AOPs varied between 40% and 55% for pad batch effluent. However, UV/O_3/H_2O_2 was found to be superior for COD removal.

Nasir et al. (2017) reported an environmentally friendly method that is both water and energy efficient for easy-care finishing by foam coating on lycra denim fabric. They applied dihydroxy ethylene urea (DHEU) resin in foam form to cotton lycra stretch denim fabric, which imparted crease recovery that was similar to conventional padded fabric. Padded and foam-coated fabric exhibited improvement of one rating in dry crocking and a half rating improvement in wet crocking as well as in the color change rating as compared to the untreated fabric. The foam technique is chemical, energy, and cost efficient as well as environmentally friendly due to formaldehyde-free resin.

Many R&D efforts (Polonca and Tavcer, 2008; Saravanan et al., 2012; Pervin et al., 2009) have been made to combine desizing, scouring, and bleaching operations to save energy and water. Harane et al. (2014) subjected gray cotton fabric to combined pretreatment and dyeing. The relative strength of these dyed samples showed good uniform dyeing results and they were very similar, or in some cases slightly higher, than conventional dyeing. The combined processes resulted in a substantial saving in water (83.17%), energy (88.60%), and time (79.3%). The single-bath pretreatment and dyeing method saves not only treatment time, but also water and energy.

Zhou et al. (2016) used a genetic algorithm for the optimization of the dyeing production schedule, to reduce the water consumption by optimization of scheduling based on dyeing color and depth. They developed a scheduling system with a database and a MATLAB program coupled with a dynamic genetic algorithm. They implemented this system in a dyeing industry, which rescheduled about 50–70 orders with unexpected orders inserted as a case study. Compared to traditional production scheduling, the optimized production schedule reduced freshwater consumption by about 18%–21%.

Yukseler et al. (2017) analyzed the BAT for wastewater from a denim manufacturing textile site. Their study focused on the selection/application of the most appropriate and sustainable in-process and end-of-pipe BAT options for the management of wastewater. Their results demonstrated that dyeing and finishing wastewaters were responsible for about 80% of the total wastewater generation, with the highest effluent load due to high COD and color. Thus, water reuse in these processes appeared to be essential in terms of water minimization. In terms of water consumption, these two processes were found to be responsible for more than 50% of the total extracted water. The BAT options suggested for the reclamation of segregated wastewater streams were nanofiltration (NF) of dyeing wastewaters with/without coagulation/microfiltration + ultra-filtration pretreatment and recovery and reuse of caustic material from the mercerization wash water.

9.5 Waterless chemical processing of textiles

The textile industry uses around 10,000 different dyes and chemicals, of which 50% are reactive dyes (Spadaro et al., 1994). Many of these chemicals are toxic and hazardous if emitted directly into bodies of water. Textile mill waste streams contain heavy metals and organics, which can lead to toxicity and eutrophication. Waterless chemical processing can be a very effective alternative to reduce toxic waste streams from the textile industry.

9.5.1 Supercritical carbon dioxide-based processing

Chemical processing of textile materials in supercritical carbon dioxide has been identified as one of the alternatives to water-based conventional processes. A supercritical fluid actually has physical properties somewhere between those of a liquid and a gas. Supercritical fluids are able to spread out along a surface more easily than a true liquid because they have lower surface tensions than liquids. At the same time, a supercritical fluid maintains a liquid's ability to dissolve substances that are soluble in the compound, which a gas cannot do. Supercritical fluids are by definition at a temperature and pressure greater than or equal to the critical temperature and pressure of the fluid. The critical pressure of CO_2 is about 1070 pounds per square inch (psi) and the critical temperature is about 31°C, so supercritical applications using CO_2 typically operate at temperatures between 32 and 49°C and pressures between 1070 and 3500 psi. A pressure-temperature (P-T) phase diagram, shown in Fig. 9.2, illustrates the nature of a supercritical fluid (Anon, 2011).

A literature survey reveals that $SCCO_2$-based processing can be adapted to the processing of polyester, for which it gives good results compared to conventional processes. There are some technical problems in adopting this technology for cotton and other natural fibers due to their hydrophilic nature. Some of the disadvantages of

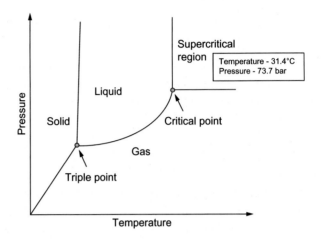

Fig. 9.2 Phase diagram of carbon dioxide (Arputharaj et al., 2015).

Table 9.10 Comparison of conventional dyeing with $SCCO_2$ (Tiwari, 2018)

Conventional dyeing	Dyeing with $SCCO_2$
Large volumes of water required for processing textiles	No water required in textile processing steps
Energy requirements are higher	Comparatively less energy required (10%–20% less)
Longer duration is required for dyeing	Reduced dyeing and washing time
Higher waste water generation	No waste water generation

supercritical dyeing include high investment cost, a complex dyeing process, which needs skilled labor, and the need to maintain a high temperature and pressure to maintain the supercritical state (Miah et al., 2013). Although there are some technical problems in the process, a comparison with the conventional dyeing process, as shown in Table 9.10, demonstrates that $SCCO_2$ has promising future prospects.

Zheng et al. (Miah et al., 2013) successfully designed a multiple industrial-scale supercritical CO_2 apparatus. They developed a polyester bobbin dyeing machine with the characteristics of an internal circulation system, and a combination dyeing process of "inner/outer dyeing" and "dynamic/static dyeing." The polyester dyeing experiments showed satisfactory dyeing with uniform and bright color. The dyed material showed excellent color fastness to washing, rubbing, and light (Zheng et al., 2016).

9.5.2 Plasma processing

Plasma is called the fourth state of matter. If a substance in its gaseous state absorbs higher energy, the outermost electrons present in the atoms will escape from the control of the nucleus and become free electrons. The atom becomes positively charged due to the loss of an electron. This chemical state of a substance is called a plasma. It consists of positive and negative ions, atoms, electrons, molecules, radicals, and photons. As the plasma is highly reactive in nature, it has the ability to react with other substances, leading to various chemical fusions and fissions. Among the different plasma production methods, atmospheric plasma processing is mostly prescribed for textile processing. It has been shown that most of the textile processing steps, such as desizing and finishing, can be carried out using plasma reactors with the less or practically no water (Kan et al., 2011; Kan and Lam, 2014; Kan and Yuen, 2006; Li and Jinjin, 2007; Man et al., 2014).

9.5.3 Zero liquid discharge

Textile processing sites generate wastewater with high salinity/TDS. Conventional "physicochemical-biological" treatment does not remove salinity in the treated effluent. The TDS content is well above the statutory limit of 2100 mg/L. Discharge of saline, but treated, wastewater pollutes ground and surface waters. ZLD refers to the installation of facilities and systems that enable the absolute recycling of permeate in industrial effluents and convert solute (dissolved organic and inorganic compounds/salts) into residue in the solid form by adopting a method of concentration and thermal evaporation. ZLD will be recognized and certified based on two broad parameters, namely, water consumption versus wastewater reused or recycled (permeate) and the corresponding solids recovered (percent total dissolved/suspended solids in the effluent). A typical ZLD system comprises the components detailed in the following subsections.

9.5.3.1 Primary treatment

This is a solid-liquid separation process involving physical processes in which the colloidal particles are precipitated. In this process, settable and floatable solids are removed. First, all the effluent is collected and mixed together completely. Then, the effluent is neutralized using coagulants in a primary clarifier. The sludge arising out of this process is removed for disposal. Color is also reduced to some extent. Finally, the pH is adjusted to 7–8 and taken to the secondary treatment process.

9.5.3.2 Secondary treatment

This is a biological process in which biodegradable organic matter is broken into simple compounds of a nonhazardous nature. This process removes BOD and dissolved and suspended organic matter by using microorganisms. Suspended solids are also removed. Organic materials are converted into carbon dioxide and sludge. In this process, microorganisms and nutrients are added into neutralized effluent coming from primary treatment in a clarifier with sufficient aeration. During this process, effluent and biological sludge containing living organisms are mixed together and aerated. The aeration is done with the help of diffusers and a blower for supplying the required oxygen for biodegradation. The aeration tank effluent is then taken into a secondary clarifier for separating the sludge. Part of the sludge is taken out for reuse. Then, the effluent is taken to a sedimentation tank where polyelectrolyte can be added for aiding flocculation of any remaining sludge.

9.5.3.3 Tertiary treatment

Effluent coming out of secondary treatment contains nonbiodegradable matter. This effluent is passed through a pressure sand filter and activated carbon filter. Finally, the effluent is passed through an RO system for recycling and reuse.

By adopting ZLD, approximately 75%–80% water can be recovered and reused. Apart from water, many useful chemicals can also be recovered. The industry only needs 25% water replenishment for its working. However, installing a ZLD system requires a huge investment compared to normal ETP. The cost of installing ZLD will be Rs. 1200–1500 per cubic meter compared to the Rs. 300–350 per cubic meter requirement of conventional ETP (Central Pollution Control Board, 2017).

9.5.4 Digital printing

Digital textile printing is described as an inkjet-based method of printing colorants onto fabric. It is used for either printing smaller designs onto garments or printing larger designs onto fabric. The latter is a growing trend in visual communication, in which advertisement and corporate branding is printed onto polyester media. Examples include flags, banners, signs, and retail graphics. With the development of a dye-sublimation printer in the early 1990s, it became possible to print with low-energy sublimation inks and high-energy disperse inks directly onto textile media, as opposed to print dye-sublimation inks on a transfer paper and, in a separate process using a heat press, transfer it to the fabric. Recent developments indicate that cellulosic fibers can also be printed using digital printing technology. The biggest advantage of digital printing is high productivity with less water usage and wastage of dyes and chemicals compared to conventional printing.

9.6 Conclusion

Textile chemical processing is very important linkage in the entire textile value chain due to its complexity and innovative nature. It is a leading consumer of water and it ranks among the top 10 water-consuming industries. Mechanical processing of textile-like spinning, weaving, and garmenting requires very little water. However, the chemical processing methods, such as scouring, bleaching, dyeing, and finishing require a substantial quantity of water. Water is mainly used for three major purposes: as a solvent for processing chemicals, as a medium for washing and rinsing, and to produce steam. After processing, the used water is released as effluent; if it is untreated, the effluent can contaminate groundwater resources. Generally, natural fibers, such as cotton and wool, require more water for processing than synthetic fibers, such as polyester and nylon. Natural fibers are hydrophilic in nature and generally water is used for the medium of dyeing and dyes are exhausted on the fabric. During processing, water is absorbed by these fibers along with dyes and chemicals. Therefore, more water is required for processing natural fibers. In contrast, synthetic fibers are generally hydrophobic and do not absorb water. For some fibers, such as polyester, the thermal fixation technique is used to exhaust the dye inside the fiber structure. Hence, efforts have to be made to reduce the water usage for processing natural fibers. The energy consumption is directly related to water consumption, as most of the energy is required to heat up process baths. Therefore, a reduction in water consumption results in a reduction in energy consumption also. Hence, reducing the use of water for processing and reusing the effluent are very important from a sustainability point of view. At present, the ZLD technique proposed by the pollution control board is the only process that can be adopted industrially for reusing treated effluent. However, the cost of installing such a setup is one of the limiting factors in implementing ZLD. The modernization of the textile processing industry with advanced machines requiring very low MLRs is one of the measures that the industry can readily take up.

The textile processing industries have been in the doldrums since the last decade of the 20th century due to environmental-related issues, especially pertaining to water. No paradigm shift has occurred in at least the last 10 years in the reduction in the usage of water, though some developments have been made in textile processing machinery. Depletion of water resources in the world, specifically in the developing countries, such as India, where many chemical processing activities are carried out, is signaling a red alarm about the future of this industry. It is predicted that if a third world war comes, it will be about water only. At present, technologies, such as supercritical CO_2 and plasma processing, look like hallucinations for textile processors. The only way out would be the mantra to reduce, recover, and reuse the water during textile processing. If the "virtual water" concept is an integral component in the whole package of integrated water resources management, issues related with water usage may be solved in the near future. New waterless technologies, such as supercritical carbon dioxide dyeing and plasma-based processing, need to be further researched for industrial adaptation.

References

Agarwal, S.C., Kumar, S., 2011. Industrial water demand in India challenges and implications for water pricing. India Infrastruct. Rep. 2011, 274–281. https://www.idfc.com/pdf/report/2011/Chp-18-Industrial-Water-Demand-in-India-Challenges.pdf.
Anon, (2011). Supercritical Carbon-Dioxide Cleaning Technology Review.
Anon (2015). Global garment and textile industries: workers, rights and working conditions, https://www.solidaritycenter.org/wp-content/uploads/2015/11/Global.Garment-Workers-Fact-Sheet.11.15.pdf.
Arputharaj, A., Raja, A.S.M., Saxena, S., 2015. Developments in Sustainable Chemical Processing of Textiles. Green fashion. Springer, Singapore, pp. 217–252.
Central Pollution Control Board, 2017. Ministry of Environment and Forests, Government of India. http://cpcb.nic.in/industry-effluent-standards/ (Accessed 13th March 2018).
Chen, L., Wang, L., Wu, X., Ding, X., 2016. A process-level water conservation and pollution control performance evaluation tool of cleaner production technology in textile industry. J. Clean. Prod. https://doi.org/10.1016/j.jclepro.2016.12.006.

Conca J (2015). Making climate change fashionable—the garment industry takes on global warming, (online) https://www.forbes.com/sites/jamesconca/2015/12/03/making-climate-change-fashionable-the-garment-industry-takes-on-global-warming/#6bbcca0879e4.

Confederation of Indian Industries, 2018, Textiles and apparels, http://www.cii.in/sectors.aspx?enc=prvePUj2bdMtgTmvPwvisYH+5EnGjyGXO9hLECvTuNtLCJDTirQ5WD1HbdWnsKry (Accessed 13th March 2018).

GG 62 (1997) Water and chemical use in the textile dyeing and finishing industry. Guide produced by the environmental technology best practice programme, UK. Available at http://www.wrap.org.uk/sites/files/wrap/GG062.pdf. (Accessed 5th May 2016).

Guyer, G.T., Nadeem, K., Dizge, N., 2016. Recycling of pad-batch washing textile wastewater through advanced oxidation processes and its reusability assessment for Turkish textile industry. J. Clean. Prod. https://doi.org/10.1016/j.jclepro.2016.08.009.

Harane, R.S., Mehra, N.R., Tayade, P.B., et al., 2014. A facile energy and water-conserving process for cotton dyeing. Int. J. Energy Environ. Eng. 5, 96. https://doi.org/10.1007/s40095-014-0096-2.

Kan C W, Lam C F, Chan C K,& Ng S P (2014) Using atmospheric pressure plasma treatment for treating grey cotton fabric, Carbohydr. Polym., 102(15) 167–173.

Kan, C.W., Yuen, C., Tsoi, W., 2011. Using atmospheric pressure plasma for enhancing the deposition of printing paste on cotton fabric for digital ink-jet printing. Cellulose 18 (3), 827–883.

Kan, C.W., Yuen, C.W.M., 2006. Low temperature plasma treatment for wool fabric. Text. Res. J. 76 (4), 309–314.

Kant, R., 2012. Textile dyeing industry an environmental hazard. Nat. Sci. 4 (1), 22–26.

Li, S., Jinjin, D., 2007. Improvement of hydrophobic properties of silk and cotton by hexa fluoro propene plasma treatment. Appl. Surf. Sci. 253 (11), 5051–5055.

Man, W.S., Kan, C.W., Ng, S.P., 2014. The use of atmospheric pressure plasma treatment on enhancing the pigment application. Vacuum 99, 7–11.

Menezes, E., Mrinal, C., 2011. Pre-treatment of textiles prior to dyeing. In: Textile Dyeing. InTech.

Miah, L., Ferdous, N., Azad, M.M., 2013. Textiles material dyeing with supercritical carbon dioxide (CO_2) without using water. Chem. Mater. Res. 3(5).

Ministry of Textiles, 2018. India, http://texmin.nic.in/sites/default/files/prod_variety_yarn.pdf

Nasir, S., Muhammad, M., Bhatti, A.A., Ahmmad, S.W., Husaain, A., 2017. Development of water and energy efficient environment friendly easy care finishing by foam coating on stretch denim fabric. J. Clean. Prod. https://doi.org/10.1016/j.jclepro.2017.03.171.

Ozturk, E., Koseoglu, H., Karaboyacı, M., Yigit, N.O., Yetis, U., Kitis, M., 2016. Minimization of water and chemical use in a cotton/polyester fabric dyeing textile mill. J. Clean. Prod. https://doi.org/10.1016/j.jclepro.2016.01.080.

Pervin, A., Davulcu, A., Eren, H.A., 2009. Enzymatic pre-treatment of cotton. Part 2: Peroxide generation in desizing liquor and bleaching. Fibres Text. East. Eur. 73 (2), 87–90.

Polonca, P., Tavcer, P.F., 2008. Bioscouring and bleaching of cotton with pectinase enzyme and peracetic acid in one bath. Color Technol. 124, 36–42.

Saravanan, D., Sivasaravanan, S., Prabhu, S., Vasanthi, M., Senthil, N.S., Raja, K., Das, A., Ramachandran, T., 2012. One step process for desizing and bleaching of cotton fabrics using the combination of amylase and glucose oxidase enzymes. J. Appl. Polym. Sci. 123, 2445–2450.

Spadaro, J.T., Isabelle, L., Renganathan, V., 1994. Hydroxyl radical mediated degradation of azo dyes: evidence for benzene generation. Environ. Sci. Technol. 28, 1389–1393.

Satish Tiwari, 2018. Dyeing of PET fibre using supercritical carbon dioxide, http://www.fibre2fashion.com/industry-article/4932/dyeing-of-pet-fibre-using-supercritical-carbon-dioxide.

Water and Related Statistics, 2010. Water Resources Information System Directorate Information System Organisation Water Planning & Projects Wing. Central Water Commission, pp. 3–9.

Yarns and fibres.com (2010) Water conservation in textile industry, http://www.yarnsandfibers.com/preferredsupplier/reports_fullstory.php?id=475&section=&p_type=General&country=Global (Accessed 13th March 2018).

Yukseler, H., Uzal, N., Sahinkaya, E., Kitis, M., Dilek, F.B., Yetis, U., 2017. Analysis of the best available techniques for wastewaters from a denim manufacturing textile mill. J. Environ. Manag. https://doi.org/10.1016/j.jenvman.2017.03.041.

Zaffalon V (2010) Climate change, carbon mitigation and textiles. Textile World, http://textileworld.com/Articles/2010/July/July_August_issue/Features/Climate_Change_Carbon_Mitigation_In_Textiles.html. (Accessed 13th March 2018).

Zheng, H., Zhang, J., Yan, J., Zheng, L., 2016. An industrial scale multiple supercritical carbon dioxide apparatus and its eco-friendly dyeing production. J. CO_2 Util. 16, 272–281.

Zhou, L., Xu, K., Cheng, X., Xu, Y., Jia, Q., 2016. Study on optimizing production scheduling for water-saving in textile dyeing industry. J. Clean. Prod. https://doi.org/10.1016/j.jclepro.2016.09.047.

Zietlow B (2013) Best available techniques in the textile sector, http://www.gpcbkp.in/live/hrdpmp/hrdpmaster/igep/content/e48745/e49028/e58164/e58169/BAT_Textile_Vortrag_GPCB_pdf.

Advances in the sustainable technologies for water conservation in textile industries

10

Luqman Jameel Rather[*,‡], Salman Jameel[†], Ovas Ahmad Dar[‡], Showkat Ali Ganie[‡], Khursheed Ahmad Bhat[†], Faqeer Mohammad[‡]
*Department of Computer Science and Engineering, University of Kashmir, Baramullah, India, [†]CSIR Bio-Organic Division, Indian Institute of Integrative Medicine, Srinagar, India, [‡]Department of Chemistry, Jamia Millia Islamia (A Central University), New Delhi, India

10.1 Introduction

Scarcity of water resources and environment pollution have become increasingly serious problems as a result of water abuse and lack of treatment, whereas global industrial water requirements keep growing at an alarming rate. The contradiction between these two trends forces industrial manufacturers to adopt cleaner production technologies to save water consumption and reduce water pollution. The textile industry is one among the high-water-use and high-water-pollution industries on the globe. Most of the wet textile processes consume large volumes of fresh water and discharge large volumes of wastewater and effluent, generally with intense color, concentrated organics, and large variations in composition ranging from inorganic finishing agents, surfactants, chlorine compounds, salts, total phosphate to polymers and organic products (Takahashi and Kumagai, 2006). The Chinese textile industrial sector, considered to be the biggest textile exporter in the world (World Trade Organization, 2016), consumes 3 billion tons of fresh water and about 2 billion tons chemical oxygen demand (COD) (National Bureau of Statistics of the People's Republic of China, 2015; World Trade Organization, 2016). For the textile industry in the European Union, which is the second largest textile exporter in the world (World Trade Organization, 2016), annual water use is 600 million m^3 (Vajnhandl and Valh, 2014). For the textile industry in Turkey, which has rapidly developed and accounts for about 35% of overall Turkish export capacity, water consumption shows a large range of values from 50 to 100 m^3/ton of finished textile (Kiran-Ciliz, 2003).

To solve the water issues in these regions, the Chinese government issued the Cleaner Production Promotion Law in 2003 (Legislative Affairs Office of the State Council of the People's Republic of China, 2003) and the Chinese textile association has issued several practical cleaner production technologies directories to textile manufacturers since then (China Dyeing and Printing Association, 2006). The European Commission has issued a reference document on the best available techniques for the textile industry (European Commission, 2002). Besides, a large number of innovative cleaner production technologies developed by academic researchers have been introduced to replace existing technologies in the textile industry. They are using natural dyes instead of synthetic dyes (Mirjalili et al., 2011; Ebrahimi and Gashti, 2015; Gashti et al., 2013; Zhang et al., 2014; Gashti and Adibzadeh, 2014), using natural additives to enhance procedures (Gashti et al., 2014; Muntaha and Khan, 2015; Gashti, 2013) and using physical production methods instead of more-polluting chemical methods, such as plasma finishing (Ebrahimi et al., 2011), ultrasonic-assisted finishing (Parvinzadeh, 2009), enzymatic treatment (Gashti et al., 2013), and waterless dyeing in supercritical carbon dioxide (Zhang et al., 2016).

Therefore, a tool for evaluating water conservation and pollution control performance of cleaner production technology is needed for textile industrial manufacturers. Thus, the authors review the most advanced methods/technologies/strategies to overcome the water scarcity issues by presenting an alternative way out to the most commonly used conventional dyeing methods. In addition, the chapter covers water management strategies and the recent advanced methods for improving dyeing procedures and reuse of wastewater.

10.2 Importance of water conservation in textiles

The textile industrial sectors are in no way different than other chemical industries in serving as a potential source of environmental pollution. Large amounts of fresh and clean water are being used in varied textile processing units. The average water use in the spinning and weaving processes is much less compared to other textile wet processing. Almost all types of dyes, auxiliary chemicals, and functional finishing agents/chemicals are applied to textile surfaces in water baths. However, water is used as a solvent for various finishing processes. Washing and rinsing are the secondary purposes of water in textile wet processes. Some water is also consumed in ion exchange, boiler, cooling water, steam drying, and cleaning.

Nowadays, researchers all around the world are thinking of water conservation due to the increased competition for clean water as water tables decline and with increasing demand from both industry and residential growth. Water and effluent costs may, in the more common cases, account for as much as 5% of production costs (Shaikh, 2009). The unnecessary use of clean and fresh water also leads to higher costs for finished textile fabrics/fibers. Huge quantities of water are being used on a daily basis for textile processing and the amounts vary depending upon the type of fabric/fiber, type of process involved, equipment used, and the nature of dyestuffs. The longer the processing sequences, the higher will be the quantity of water required. At the end of textile finishing processes, large amounts of water are being used for washing. However, a wide variation in water consumption is observed due to the use of old and new technologies and the differences in processing steps. Every textile processor should have knowledge of the quantity of water used for processing.

Water conservation and reuse are becoming a necessity for the textile industry and this has led researchers to develop new alternative ways for decreasing the costs of wastewater treatment and to lower the rates used for finished textile products;

nowadays, this is becoming one of the highlighted research areas. Avoiding discharge violations with proper disposal of effluents results in a significant inducement for water conservation and reuse. A schematic representation of water use during different stages of dyeing and finishing of textile fabric/fiber is shown in Fig. 10.1. The implementation of water conservation and reuse programs/technologies/strategies will keep treatment facilities on hold, and may divert the available funds for expansion or improvements in other processes and equipment. However, a site survey should be conducted for developing effective water conservation and reuse programs. It would be extremely helpful to develop a spreadsheet and/or diagram of the water usage.

Water conservation significantly reduces effluent volume. A water conservation program can cut water consumption by up to 30% or more and the cost savings can pay for the required materials in a very short time (Balachandran, 2008). Because the average plant has a large number of washers, the savings can add up to thousands of rupees per year. Other reasons for large effluent volumes are the choice of inefficient washing equipment, excessively long washing circles, and use of fresh water at all points of water use. The equipment used in a water conservation program is relatively inexpensive, consisting in most cases of valves, piping, small pumps, and tanks only. The operating costs for these systems are generally very low. Routine maintenance and, in some cases, electricity for the pumps, would be the major cost components. The payback period for a water conservation system will vary with the quantity of water saved, sewer fees, and costs for raw water and wastewater treatment (US Environmental Protection Agency). In addition to the direct cost savings, a water conservation program can reduce the capital costs of any required end-of-pipe wastewater treatment system. Personnel from the textile industry need to be aware of the potential of water conservation so that they can help their organization realize the benefits. In general, water conservation measures lead to:

a. Reduction in processing cost.
b. Reduction in wastewater treatment cost.
c. Reduction in thermal energy consumption.
d. Reduction in electrical energy consumption.
e. Reduction in pollutants load.

Fig. 10.1 Water usage balance in dyeing and finishing of textile materials (Balachandran, 2008).

f. Water reuse measures reduce hydraulic loadings to treatment systems by using the same water in more than one process. Water reuse resulting from advanced wastewater treatment (recycle) is not considered an in-plant control because it does not reduce hydraulic or pollutant loadings on the treatment plant. Examples of process water reuse are listed in the following sections.

10.2.1 Reuse of water jet weaving wastewater

The jet weaving wastewater can be reused within the jet looms. Alternatively, it can be reused in the desizing or scouring processes, provided that in-line filters remove fabric impurities and oils.

10.2.2 Reuse of bleach bath cotton

Cotton blend preparations are performed using continuous or batch processes and usually are the largest water consumers in a mill. Continuous processes are much easier to adapt to wastewater recycling/reuse because the waste stream is continuous, shows fairly constant characteristics, and usually is easy to segregate from other waste streams. Waste stream reuse in a typical bleach unit for polyester/cotton blend and 100% cotton fabrics would include recycling j-box and kier drain wastewater to saturators, recycling continuous scour wash water to batch scouring, recycling washer water to equipment and facility cleaning, reusing scour rinses for desizing, and reusing mercerized wash water or bleach wash water for scouring. Preparation chemicals, however, must be selected in such a way that reuse does not create quality problems, such as spotting. Batch scouring and bleaching are less easy to adapt to recycling of waste streams because streams occur intermittently and are not easily segregated. With appropriate holding tanks, however, bleach bath reuse can be practiced in a similar manner to dye bath reuse and several pieces of equipment are now available that have the necessary holding tanks.

10.2.3 Reuse of final rinse water from dyeing for dye bath

The rinse water from the final rinse in a batch dyeing operation is fairly clean and can be used directly for further rinsing or to make up subsequent dye baths. Several woven fabric and carpet mills use this rinse water for dye bath preparation.

10.2.4 Reuse of soaper wastewater

The colored wastewater from the soaping operation can be reused at the back gray washer, which does not require water of a very high quality. Alternatively, the wastewater can be used for cleaning floors and equipment in the print and color shop.

10.2.5 Reuse of dye liquors

The feasibility of dye liquor reuse depends on the nature of the dye used and the shade required on the fabric or yarn as well as the type of process involved. It has already been applied for disperse dyeing polyester, reactive dyeing cotton, acid dyeing nylon, and basic dyeing acrylic on a wide variety of machines. However, having many dyeing processes in which the shades required are much more varied and unpredictable would make reuse difficult. But, under the right conditions, dye liquor could be reused up to 10 times before the level of impurities limits further use.

10.2.6 Reuse of cooling water

Cooling water that does not come in contact with fabric or process chemicals can be collected and reused directly. Examples include condenser-cooling water, water from water-cooled bearings, heat-exchanger water, and water recovered from cooling rolls, yarn dryers, pressure dyeing machines, and air compressors. This water can be pumped to hot water storage tanks for reuse in operations, such as dyeing, bleaching, rinsing, and cleaning, for which heated water is required or used as feeding water for a boiler.

10.2.7 Reusing wash water

The most popular and successful strategy applied for reusing wash water is counter-current washing. The counter-current washing method is relatively straightforward and inexpensive. For both water and energy savings, counter-current washing is employed frequently on continuous preparation and dye ranges. Clean water enters at the final wash box and flows counter to the movement of the fabric through the wash boxes. With this method, the least contaminated water from the final wash is reused for the next-to-last wash and so on until the water reaches the first wash stage, where it is finally discharged. Direct counter-current washing is now generally built into the process flow sheet of new textile mills. It is also easy to implement in existing mills where there is a synchronous processing operation. Washing and rinsing are both important for reducing impurity levels in the fabric to predetermined levels. Water and wastewater treatment prices are increasing; the optimization of water use pays dividends. One possible option is to reduce rinse water use for lighter shades. Here are some successful water reduction projects in batch and continuous operations.

a. *Winch dyeing*: 25% reduction will occur by avoiding overflow rinsing.
b. *High and low*: Replacing the overflow with pressure jet dyeing batch-wise rinsing can cut water consumption by approximately 50%.
c. *Beam dyeing*: About 60% of water could be saved by preventing overflows during soaking and rinsing. Automatic controls proved to be quite economical with a payback period of about 4 months.
d. *Jig dyeing*: A wide range of reductions, ranging from 15% to 79%, is possible by switching from the practice of overflow to stepwise rinsing. Rinsing using a spray technique is also effective.

e. *Cheese dyeing*: A reduction of around 70% is possible following intermittent rinsing.
f. *Continuous operation*: A 20%–30% saving was realized by introducing automatic water stops. Counter-current washing proved to be the most effective method. Horizontal washing equipment delivered the same performance as two vertical washing machines, using the same amount of water.

10.3 Sustainable strategies for water conservation in textiles

Decoupling of economic growth between water consumption and wastewater discharge has laid the foundation (principle) for achieving sustainable development in water conservation. Lot of research work has been done and is currently ongoing regarding the development of water conservation at textile mills. Some of the techniques/strategies that are being used by a wide variety of mills are discussed in detail (Chen et al., 2015a, b; Zhu et al., 2013). However, a reduction in water consumption (10%–30%) can be achieved by taking strict housekeeping measures. Good housekeeping measures/strategies are often carried out without significant investments, leading to water conservation in a cost-effective manner along with minimization of the energy and chemicals. A simple examination can uncover wastewater in the form of:

a. Hoses left running.
b. Broken or missing valves.
c. Excessive water use in washing operations.
d. Leaks from pipes, valves, and pumps.
e. Cooling water or wash boxes left running when machinery is shut down.
f. Defective toilets and water coolers.

Implementation of the cost-effective strategies listed above may lead to a greener environment with least wastage of pure drinkable water and, subsequently, less wastewater. In this part of the chapter, the authors have reviewed the most important, eco-friendly, and ecologically benign water conservation technologies.

10.3.1 Decoupling of water conservation by water footprint method

Decoupling analysis has received great attention worldwide with respect to the relation of economic growth and growing environmental pressures (Fischer-Kowalski et al., 2016). Implementation of green economic development requires a breakdown of the link between the economy and environmental hazards, which is a basic principle of human development (Organization for Economic Co-operation and Development, 2001, 2002). Recent research studies have been focusing on the decoupling between economic growth and industrial water usage/generation of wastewater (Caneghem, 2010; Chen et al., 2015a, b; Conrad and Cassar, 2014; Gilmont, 2014; Lei and Su, 2015; Wu and Wang, 2013; Zhu et al., 2013). However, an integrated improvement of water management at the industrial level requires a reduction in water consumption and wastewater discharge simultaneously.

The water footprint method provides an easy and industrially feasible way to integrate water resources and water environmental science as one factor through performing an accurate decoupling analysis. Wang et al. (2013b, b) applied water footprint methods in China's textile industry. Additionally, a case study was performed by Zhang and Yang, 2014 in Heilongjiang Province of China by integrating the water footprint method for decoupling of agricultural output, water consumption, and the environmental impact of crop production. Decoupling of the WF of the food industry from social development was performed in the food industries of the Middle East and North African countries in 1961–2009 (Gilmont, 2015). Pan and Chen (2014) represent water resource utilization in terms of WF and they investigated the decoupling relationship between the footprint and economic growth. A weakening decoupling trend was found between WF and GDP growth.

10.3.2 Foam technology

The world's most important problem nowadays is the shortage of pure drinking water. Conventional textile processing requires large amounts of water to dissolve and apply the required chemicals onto the fabric surfaces. Hence, the amount of available drinking water is reduced and enormous water effluents are generated. However, modern day technologies during textile manufacturing are responsible for making the processes more energy intensive. Padding is the conventional wet processing technique in which chemicals are applied onto the textile material surfaces. Aqueous solutions of active chemicals (1%–8%) are applied through the padder to squeeze the excess liquor in order to achieve wet pickup of 60%–90% (Cook, 1994). However, the excessive use of water is responsible for an increase in the processing cost of the padding process with a reduction in the volumes of available drinking water and an increment in the generation of wastewater effluents. Presently, it has been seen that not all the organizations are purifying contaminated wastewater properly, which after consumption cause various diseases, including cancer. On the other hand, increasing awareness among people of the impact of the textile industries and their adverse effects on the environment has led to the formulation of strict legislation all over the world. Energy efficiency, environmental friendliness, and low water consumption have become prerequisite principles for modifying the conventional textile processes. Hence, textile manufacturers have focused their efforts and developed a new unconventional foam coating technology that opened a new era of water and energy conservation in textile wet processing (Ramachandran et al., 2008).

Foam application technology can only use extremely low wet pickup of about 10%–20% of the desired recipe as compared to 60%–90% of the conventional wet pickup. Furthermore, a water-based recipe is converted to 7–10 times the volume of foam, consequently, higher weights of the fabric can be treated with lesser amounts of chemicals. Foam increases the surface area of the chemicals, which results in a subsequent reduction in the liquor required as compared to the conventional padding technique. Consequently, relatively low amounts of chemicals and heat are required to finish the fabric. A higher production speed, reduced curing time, and significantly lower effluent targets are also achieved. It has been claimed that foam technology can lead to a reduction in water consumption of up to 80% and a reduction in energy consumption of up to 65% as compared to the conventional padding processes (Hasanbeigi and Price, 2015). Therefore, foam technology is becoming increasingly popular in the textile industry. Foam technology can be equally beneficial for various wet processing processes, such as yarn sizing, mercerization, dyeing, wrinkle resist finishing, fire retardant finishing, and oil and water repellent finishing of the textile (Li et al., 2014; Gregorian et al., 1983; Juanjuan et al., 2015; Wadsworth and Wey, 1988; Zhu et al., 2015). The superiority of the foam over padding is seen in lower water use and low pickup, which leads to an increase in the production speed and lower energy consumption (Ramachandran et al., 2008). Researchers have reported comparable results in the performance parameters, such as ease of care and fire resistance, using foam instead of padding, with minimal effect on the physical properties. An extended use of heat to evaporate excess water leads to damage in the physical properties of the fabric (Nasr et al., 2016; Chin-kuen and Chi-wai, 2016). The foam application technique has been reported for the application of various toxic resins, such as a high level of formaldehyde chemicals, formaldehyde containing the most effective dimethylol dihydroxy ethylene urea (DMDHEU)-based resin, and the uniform distribution of DMDHEU and consequent release of a high amount of formaldehyde (Gonzales and Reinhardt, 1986, 1987; Rowland et al., 1983). However, there is a need to use alternative formaldehyde-free resin, which can impart the desired results with the foam technique. In addition, the application of the foam technique for stretch denim fabric along with the use of zero formaldehyde cross-linker have not been reported in the literature.

10.3.3 Water management and recycling through advanced oxidation processes

For the past 20–30 years, the European framework has been policy driven in several directions to minimize pollution. Among them, the most significant directions for industries were the integrated pollution prevention and control (IPPC) directive, revoked by the industrial emission directive (IED) and water framework directive (Evrard et al., 2016). It assimilates the basic rules of environmental protection to the core industrial manufacturing process for resource optimization and pollution prevention, or cleaner production (Kocabas et al., 2009). Cleaner production and pollution prevention/minimization tactics started to take place in environmental protection policies and regulations of the countries. One mean of achieving the purpose is to execute techniques called best available techniques (BAT) to find a suitable balance between environmental performance and technical/economical convenience (Laforest, 2014). The principle of Turkish BREF is to minimize the unfavorable effects of the textile industry, control discharges to collecting media (water, air, and soil), and effectual consumption of raw materials and energy by using cleaner production techniques. For water consumption, the textile industry ranks second among all the industries in Turkey and is responsible for solid waste, dust pollution, noise pollution, and bulk water pollution (Ozturk et al., 2009). The production processes of the textile industry can be categorized as wet processes (finishing, dyeing, washing, etc.) and dry processes (knitting, weaving, fixing, etc.). Wastewater streams that arise from wet processing contaminate fresh water with spent chemicals and dyestuffs, which comprise large amount of organic matter, color, and salts. In addition to wastewater production, fresh water utilization is also a thoughtful concern, particularly in those countries experiencing water shortages or those that will be in the near future. Fresh water utilization and wastewater discharge are both treated with national discharge limits. Fresh water and effluent treatment, in the common cases, may cost as much as 5% of the production costs. The textile industry utilizes a huge amount of water to produce 1 kg of finished fabric. Fiber production, spinning, and weaving utilize much less water compared with wet processing. According to the United States Environmental Protection Agency (US-EPA), a unit manufacturing 20,000 lbs/day of fabric uses 36,000 L of water in wet processing. The amount of water needed to produce one unit of fabric differs according to the fabric quality, shade required, dyes and chemical used, liquor ratio and process conditions (Brik et al., 2006).

According to various investigators, the average water consumed to manufacture 1 kg of fabric varies between 200 and 400 L (Amar et al., 2009; Brik et al., 2006). Evaluations have shown that each year 280,000 tons of dyes are released to the collecting water bodies with 8000 dyes and 6900 auxiliary chemicals and salts (Asghar et al., 2015; Hazrat et al., 2013). It has also been found that 1 kg of chemical and auxiliaries are utilized for preparing 1 kg fabric (Vajnhandl and Valh, 2014). A huge quantity of salts, along with dyes and other chemicals, are the main concern of textile wet processing.

Wet processes in the textile industry need water of the best quality regarding the content of dyes, detergents, and suspended solids. Therefore, a purification treatment to reuse water must perform reliably to enable its use instead of simple discharge according to the limits imposed by legislation. In order to meet legislative requirements, textile wastewaters are generally treated in a chemical-physical, or usually in an activated sludge biochemical plant. In order to have water that can be reused in production cycles, especially dyeing processes, water requires further treatment (Klose, 1993). Encouraging results have also been reported regarding the application of ozone treatment for the cleansing of textile wastewater for reuse (Ciardelli and Ranieri, 1998). This research background has been recently converted into an awareness on the part of industrial plants for the full employment of the technique. It is fascinating to watch if this approach can be linked to the application of membrane technology for a more effectual management and reprocessing of textile dyeing and fulling plant water resources. The approaches based on an ozonization plant for bioresistant pollutant oxidation and ultrafiltration and reverse osmosis membrane treatment may be compared to the present management of wastewater in the dyeing and finishing textile plant. Curiosity in membrane processing and other industrial applications is increasing rapidly due to the latest technological innovations that make them trustworthy and economically achievable with respect to other alternative systems. The purpose of the effluent pretreatment step (coagulation, sand filtration, disinfection) is vital to guarantee a decent and constant performance in the membrane efficiency (Coste and Marmagne, 1996).

Among the wastewater treatment technologies, AOPs are very applicable for the removal of color, refractory organic compounds, and particularly chromophoric structures. These processes are mostly chemical oxidation processes that consume one of the strongest oxidizing species, namely, hydroxyl radicals (HO·). They insistently and unselectively strike most refractory compounds, which may cause complete mineralization at the end (Ciardelli and Ranieri, 2001). Once produced, the hydroxyl radicals may strike organic chemicals by radical addition (Eq. 10.1), hydrogen abstraction (Eq. 10.2), and electron transfer (Eq. 10.3).

$$R + HO^{\bullet} \rightarrow ROH \tag{10.1}$$

$$R + HO^{\bullet} \rightarrow R^{\bullet} + H_2O \tag{10.2}$$

$$R^n + HO^{\bullet} \rightarrow R^{n-1} + OH^{-} \tag{10.3}$$

Generally, AOPs use a combination of strong oxidizing agents (e.g., H_2O_2, O_3) with catalysts (e.g., transition metal ions) in the presence of irradiation (e.g., ultraviolet, visible, ultrasound). Among several AOPs yielding hydroxyl radicals, hydrogen peroxide/UV light process, Fenton's reactions, and ozonation at high pH appear to be some of the most common technologies for wastewater treatment as revealed by the huge amount of data available in the literature (Asghar et al., 2015; Cardoso et al., 2016). AOPs have numerous advantages over conventional treatments, as well as the ability to control fluctuating flow rates and compositions, rapid reaction rates, and the absence of secondary wastes. AOPs also have the capacity to remove wastewater spoiled with refractory compounds, rather than collecting or transferring pollutants into another phase, though, advanced oxidation processes generally have better capital and operating costs associated with biological treatment. A cleaner production approach in the textile industry, with the concept of BAT, is challenging and should be supplemented by regulatory, environmental, and technical constraints.

10.3.4 Process-level water conservation and pollution control method

A literature review of the present evaluation tools connected to water issues in the textile industry has been conducted. These tools are classified into two categories. The first category measures the water use and water pollution of specific objects at a time. The second category measures the water conservation and pollution control performance of particular technologies or methods before and after their application for a period of time. For the first category, the 7th evaluation tool, developed by Chen et al. (2015a, b), comprises water withdrawal, water consumption with significance of water reuse, and water assimilation, while other tools (Chapagain et al., 2006; Wang et al., 2013b, b; Yan et al., 2014; Joa et al., 2014; Zhang et al., 2016) only measure one or two of these three impacts. By assessing both water withdrawal and water consumption, reflecting the effect of water reuse and assessing the level of water pollution at the same time, the 7th tool may obtain the hotspots with high water use and high water pollution in a more comprehensive way. However, this tool can only measure the impacts at one time instead of showing the performance improvement.

For the second category, which emphasizes the measurement of performance improvement, three drawbacks were found:

a. Some tools only concentrate on either water conservation or pollution control. However, water conservation performance and pollution control performance are uniformly important for cleaner industrial production technologies because some industrial manufacturers decrease the pollution concentration by removing a huge amount of fresh water to dilute the effluent (Piatti et al., 2011; Gu et al., 2016).
b. Other tools, which reflect both water conservation and pollution control, only pay attention on the reduction of water withdrawal but ignore changes in water consumption, which may give rise to overestimation of the performance (Guo et al., 2006; Alberto et al., 2012; Emrah et al., 2015).
c. All of the second-category tools do not measure how technologies bring about performance improvements in the various production processes in a plant. The textile industry has a vast variety of production processes. Without assessment at the process level, textile manufacturers cannot comprehend the operation mechanism of technologies and are not able to attain specific improvement methods on various processes for increasing performance.

From the comprehensiveness and the process-level data requirement of the first-category evaluation tools, a new process-level water conservation and pollution control performance evaluation tool for cleaner production technology was developed for the textile industry and this performance evaluation tool and the other two existing evaluation tools were applied to the textile industry to measure the water conservation and pollution control performance of a cleaner production technology that is extensively used in textile dyeing and printing.

10.3.5 Physical separation technologies

10.3.5.1 Membrane separation

Membranes are made of numerous materials that can be of liquid or solid, natural, or synthetic origin. There may be inorganic (ceramic) or organic (polymeric) materials present in these membranes. Polymeric membranes (cellulose acetate, polysulfone, polyamide, polyvinyldene fluoride) appear to be most significant for their applications in the field of textile wastewaters. The state of the art in the field of membranes is presently anisotrope (asymmetric) membranes. Anisotrope membranes have a thin film that prevents the entrapment of suspended solids into the membrane body and are thus less subject to aging and flow reduction than symmetric membranes. One of the latest developments in composite membranes involves a thin film with small pores that is laid on a classical asymmetric membrane. This kind of membrane, originally grown for reverse osmosis, is presently gaining application in ultrafiltration.

Membrane configurations are generally categorized depending on the types of modules adopted. The most common module is the spiral-wound module. The cylindrical form of this module encloses the membrane in itself with a net that prevents membrane-to-membrane contact and allows the feed to flow. The flow of the concentrate goes parallel to the axis of the membrane module while the permeate flows in the radial direction (cross-flow) through the membrane, reaches the collector, and then flows axially in a separate circuit. One of the most frequent problems in membrane plants, also in the textile field, is the progressive deterioration of the quality of the permeate produced. The flow reduction has to be attributed to a reversible (concentration polarization) or irreversible (fouling) increased resistance of the membrane to the permeate flow. Membrane disinfection is necessary to prevent biofouling of the membrane surface but reverse osmosis is not generally resistant to the more common chlorine-based disinfecting agents.

Membrane technology has established numerous industrial applications, supported by sufficient literature references, particularly for ultrafiltration. The most significant ones involve the treatment of tannery and textile wastes, oily emulsions, and electrophoretic painting (Denaro, 1993). Membrane processes have been investigated for the treatment and reuse of effluents mostly from textile dyebaths (Drioli, 1992; Buckley, 1992). Ultrafiltration permits water clarification and disinfection, without byproducts, in a single step and with a uniform permeates quality. At pressures ranging from 2 to 10 bar particles of dimension greater than 1 nm are separated by the ultrafiltration process, which removes bacteria, viruses, proteins and some sugars from the effluent, and regrowth may not occur after treatment (Gadani et al., 1996). The removals of salts from solutions are separated by reverse osmosis, which is another membrane process (Marinas, 1991).

10.3.5.2 Nanofiltration process

The use of nanofiltration (NF) membranes is one of the most recent and widely accepted wastewater treatment processes for the separation of low-molecular-weight compounds (200–1000 g/mol) (Akbari et al., 2001) and divalent metal salts (Marucci et al., 2001). Reuse of permeate water in preparation of the dye bath is possible and emerging technologies, such as NF membranes, reject dyes and other organic molecules, while NaCl and other monovalent salts pass through the membrane. Applying a membrane potential and the sieving mechanism can hinder the transport of different component through the membrane (Akbari et al., 2001; Marucci et al., 2001). Previously, several researchers have performed theoretical evaluations of the salt rejection description of NF membranes in dye mixtures based on the Spiegler and Kedem model (Gilron et al., 2001; Koyuncu and Topacik, 2002; Levenstein et al., 1996; Perry and Linder, 1989; Spiegler and Kedem, 1966; Xu and Spencer, 1997). The latest and widely accepted model is the Koyuncu and Topacik (K–T) model (Koyuncu and Topacik, 2002), which describes the movement of dye components and salt mixtures on the basis of concentration polarization effects. According to the Koyuncu and Topacik (K–T) model, the observed salt rejection is given as follows:

$$R_{sOBS} = 1 - (1 - R_0)(\alpha) \left(1 + \frac{\upsilon c_{Dm}}{\alpha c_{sm}} \exp\left(\frac{J_\upsilon}{k_D}\right)\right)^{0.5} \tag{10.4}$$

where R_{sOBS} is the observed salt rejection in the presence of an organic ion, R_0 is the real single salt rejection, v is the number of charges of the organic ion, c_{sm} and c_{Dm} are the feed salt and organic ion concentrations, α is the gel polarization effect on salt removal, and k_D is the mass transport coefficient of dye.

However, the K–T model does not evaluate the effects of auxiliary chemicals in reactive dyeing. Koyuncu et al. (2004) studied the recycling of salt (NaCl, NaOH, Na_2CO_3) and permeate reuse in reactive dye bath wastewaters supplied by the local cotton textile industry by using NF membranes and studied the economic implications based on experimental results. A schematic representation of this kind of application is given in Fig. 10.2.

Water, NaCl, and other salts pass the membrane, whereas dye molecules are rejected. Permeate water cannot be used directly as different types of reactive dye baths possess different NaCl concentrations. However, permeate salt concentration should be measured to make the process easy and rapid. For enhancing the process, the permeate water should be diluted if the NaCl concentration is lower (Koyuncu et al., 2004). From the experimental results of Koyuncu et al. (2004), it was found that a high color rejection, greater than 99.9%, was achieved. Increased pressure led to increased rejection levels, indicating the need for optimization to be done between increasing pressure and required values of recovered water, salt, and color rejections. In addition, the choices of NF membranes are very important, depending upon the type of dyes and nature of the salts (Van der Bruggen et al., 2001). The ideal NF membrane is optimized on the basis of flux value, color, and salt rejection.

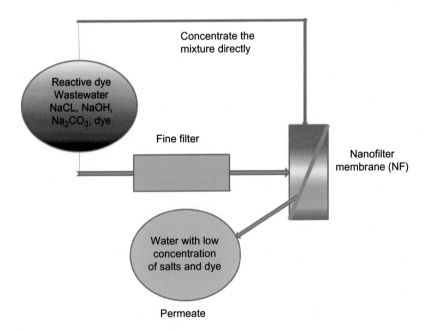

Fig. 10.2 A general Schematic representation of nanofiltration membranes for the treatment of reactive dye house wastewaters (Koyuncu et al., 2004).

10.4 Challenges and opportunities for textiles and fashion

Water reuse is a growing field and many projects have been occurring throughout Europe in the last 15 years. Most of the northern EU countries have abundant water resources. In this case, the need for extra supply through the reuse of treated wastewater is not a priority, but the protection of the receiving environment is considered to be an important issue. The situation is different in the southern EU countries, in which the additional resource brought in by water reuse has promoted the implementation of a number of new projects. One of the major constraints for water reuse and its public acceptance is the lack of relevant legislation at the EU level. As a result of this situation, both strict and flexible standards can be found in Europe, even in the same country (Spain, for example), illustrating an important equity issue, which needs to be addressed. Water shortages, particularly during periods of drought, have necessitated stricter control measures on rates of water consumption and development of alternative water sources (Asano, 1998). Advances in the effectiveness and reliability of wastewater treatment technologies have improved the capacity to produce recycled water that can serve as an alternative water source in addition to meeting water quality protection and pollution abatement requirements (Lazarova, 2000). In developing countries, particularly those in arid parts of the world, reliable low-cost technologies (both for treatment and reuse) are needed for acquiring new water supplies and protecting existing water sources from pollution. The implementation of water reuse promotes the preservation of limited water resources in conjunction with water conservation and watershed protection programs (Asano, 1998).

Additionally, companies in non-European countries, such as Turkish textile companies, have to face water shortage problems. Therefore, companies have to look for alternatives (improving wastewater treatment systems) to meet their water requirements. Closing the water loop is the ultimate solution for achieving sustainability in industrial water consumption, preventing water pollution and sustaining the ground water reserves for the future. There are many problems that are thought to be hurdles in closing this loop and reusing process water for technological and other purposes. Therefore, a combination of the most suitable options should be tailored to meet the future industrial requirements for 100% water recycling for technological purposes.

Therefore, a common approach used to evaluate water scarcity is the exploitation rate of water resources (the ratio between the volume of available renewable water resources and annual withdrawals). When the exploitation rate exceeds 20% of existing reserves, water management becomes a vital element in a country's economy. These data suggest that to meet future needs, many countries will have to manage water resources far more efficiently than they do now. In view of the current outlook for the use of water resources over the whole EU area, existing policies need to be reoriented toward a better integrated water management while minimizing health and environmental risks. The New European Water Framework Directive will lay the foundation for such an approach, including river basin management and water quality objectives. This should translate to a better control of polluting discharges over the long term.

In several countries, hydrological plans have been drawn up (Spain and France) or are in the process of being drawn up (Greece and Portugal). These plans can be effective tools for action but the existing ones:

a. Do not include integrated water resource schemes.
b. Are dominated by the significance of short-term requirements.
c. Are still mostly tuned toward increasing water availability than toward better management of the water demand in spite of recent efforts to address these new challenges.

As a result, the reuse of treated wastewater could become an important water management option, both to shore up conventional resources and to reduce the environmental impact of discharges. Such reuse is foreseen within the water master plans of several countries and is already being done. However, a number of technical and regulatory issues remain to be addressed to make sure there are no undesirable impacts on the environment or on public health. In addition, water reuse operations still require better control and appropriate staff training.

10.5 Conclusion

Recycled water is a reliable source of water that must be taken into account in formulating a sustainable water policy. A major catalyst for the evolution of water reuse has been the need to provide alternative water sources to satisfy water requirements for irrigation, industry, and urban nonpotable and potable water applications due to unprecedented growth and development in many regions of the world. The concept of deriving beneficial uses from treated municipal and industrial wastewater coupled with increasing pressures on water resources has prompted the emergence of water reuse as an integral component of water resource management. The inherent benefits associated with the recycling of treated wastewater instead of discharging it include preservation of higher quality water resources, environmental protection, and economic advantages.

Acknowledgments

Financial support provided by University Grants Commission, Govt. of India; New Delhi through UGC non-NET fellowship for *OAD*, SRF for *SAG* and CSIR-Gate fellowship for *SJ* is highly acknowledged.

References

Akbari, A., Desclaux, S., Remigy, J.C., Aptel, P., 2001. Treatment of textile dye effluents using a new photografted nanofiltration membrane. Desalination 149, 101–107.

Alberto, B., Hugo, V., Marco, C., Nora, S., 2012. Evaluation of the environmental impacts of a cleaner production agreement by frozen fish facilities in the Biobío region, Chile. J. Clean. Prod. 26, 95–100.

Amar, N.B., Kechaou, N., Palmeri, J., Deratani, A., Sghaier, A., 2009. Comparison of tertiary treatment by nanofiltration and reverse osmosis for water reuse in denim textile industry. J. Hazard. Mater. 170, 111–117.

Asano, T. (Ed.), 1998. Wastewater Reclamation and Reuse. Water Quality Management Library. In: vol. 10. Technomic Publishing Inc., Lancaster, PA.

Asghar, A., Raman, A.A.A., Daud, W.M.A.W., 2015. Advanced oxidation processes for in-situ production of hydrogen peroxide/hydroxyl radical for textile wastewater treatment: a review. J. Clean. Prod. 87, 826–838.

Balachandran, S., 2008. Efficient water utilisation in textile wet processing. IE(I)-TX 89, 26–29.

Brik, M., Schoeberl, P., Chamam, B., Braun, R., Fuchs, W., 2006. Advanced treatment of textile wastewater towards reuse using a membrane bioreactor. Process Biochem. 41, 1751–1757.

Buckley, C.A., 1992. Membrane technology for the treatment of dyehouse effluents. Water Sci. Technol. 25, 203–209.

Caneghem, J.V., 2010. Eco-efficiency trends of the Flemish industry: decoupling of environmental impact from economic growth. J. Clean. Prod. 18, 1349–1357.

Cardoso, J.C., Bessegato, G.G., Zanoni, M.V.B., 2016. Efficiency comparison of ozonation, photolysis, photocatalysis and photoelectrocatalysis methods in real textile wastewater decolorisation. Water Res. 98, 39–46.

Chapagain, A.K., Hoekstra, A.Y., Savenije, H.H.G., Gautam, R., 2006. The water footprint of cotton consumption an assessment of the impact of worldwide consumption of cotton products on the water resources in the cotton producing countries. Ecol. Econ. 60, 186–203.

Chen, L.Z., Ding, X.M., Wu, X.Y., 2015a. Water management tool of industrial product a case study of screen printing fabric and digital printing fabric. Ecol. Indic. 58, 86–94.

Chen, Q., Li, J.M., Management, S.O., 2015b. Analysis on the decoupling relationship between marine environmental stress and marine economic growth in China. Mar. Environ. Sci. 34, 827–833.

China Dyeing and Printing Association, 2006. The Clean Production Technology Directory of Key Industries in China. Available from: http://www.cdpa.org.cn/zxzx/zcfg/7029.htm.

Chin-kuen, P., Chi-wai, K., 2016. Relationship between curing temperature and low stress mechanical properties of titanium dioxide catalyzed flame retardant finished cotton fabric. J. Fib. Poly. 17 (3), 380–388. https://doi.org/10.1007/s12221-016-5809-0.

Ciardelli, G., Ranieri, N., 1998. Water Recycling in the Textile Industry: Several Case Studies. Publishing Co. Inc., pp. 171–198.

Ciardelli, G., Ranieri, N., 2001. The treatment and reuse of wastewater in the textile industry by means of ozonation and electroflocculation. Water Res. 35, 567–572.

Conrad, E., Cassar, L.F., 2014. Decoupling economic growth and environmental degradation: reviewing progress to date in the Small Island State of Malta. Sustainability 6, 6729–6750.

Cook, F.L., 1994. Less is more in applying chemicals to textiles. Tex. World 60–65.

Coste, C., Jacquart, J.C., Marmagne, O., 1996. Effluents decoloration. L'Ind Text. 1278, 46–51.

Denaro, R., 1993. Ultrafiltration treatments of oily emulsions: models of calculation and industrial applications. Ing. Ambient. 22, 259–270.

Drioli, E., 1992. Membrane operations for the rationalization of industrial productions. Water Sci. Technol. 25, 107–125.

Ebrahimi, I., Gashti, M.P., 2015. Extraction of juglone from Pterocarya fraxinifolia leaves for dyeing, anti-fungal finishing, and solar UV protection of wool. Color. Technol. 131 (6), 451–457.

Ebrahimi, I., Kiumarsi, A., Gashti, M.P., Rashidian, R., Norouzi, M.H., 2011. Atmospheric-air plasma enhances coating of different lubricating agents on polyester fiber. Eur. Phys. J. Appl. Phys. 56 (1), 10801.

Emrah, O., Mustafa, K., Ulku, Y., Nevzat, O.Y., Mehmet, K., 2015. Evaluation of integrated pollution prevention control in a textile fiber production and dyeing mill. J. Clean. Prod. 88, 116–124.

European Commission, 2002. Integrated Pollution Prevention and Control – Reference document on Best Available Techniques for the Textiles Industry. In: Directorate-General Joint Research Centre, Institute for Prospective Technological Studies (Sevilla) Technologies for Sustainable Development. Bureau, European IPPC.

Evrard, D., Laforest, V., Villot, J., Gaucher, R., 2016. Best available technique assessment methods: a literature review from sector to installation level. J. Clean. Prod. 121, 72–83.

Fischer-Kowalski, M.; Swilling, M.; von Weizsäcker, E.U.; Ren, Y.; Moriguchi, Y.; Crane, W.; Krausmann, F.; Eisenmenger, N.; Giljum, S.; Hennicke, P., et al. 2016. Decoupling Natural Resource Use and Environmental Impacts from Economic Growth. A Report of the Working Group on Decoupling to the International Resource Panel. http:/www.unep.org/resourcepanel/decoupling/files/pdf/decouplingreport.

Gadani, V., Irwin, R., Mandra, V., 1996. Ultrafiltration as a tertiary treatment: joint research program on membranes. Desalination 106, 47–53.

Gashti, M.P., 2013. Effect of colloidal dispersion of clay on some properties of wool fiber. J. Dispers. Sci. Technol. 34 (6), 853–858.

Gashti, M.P., Adibzadeh, H., 2014. Ultrasound for efficient emulsification and uniform coating of an anionic lubricant on cotton. Fibers Polym. 15 (1), 65–70.

Gashti, M.P., Katozian, B., Shaver, M., Kiumarsi, A., 2014. Clay nanoadsorbent as an environmentally friendly substitute for mordants in the natural dyeing of carpet piles. Color. Technol. 130 (1), 54–61.

Gashti, M.P., Rashidian, R., Almasian, A., Zohouri, A.B., 2013. A novel method for colouration of cotton using clay nano-adsorbent treatment. Pigm. Resin Technol. 42 (3), 175–185.

Gilmont, M., 2014. Decoupling dependence on natural water: reflexivity in the regulation and allocation of water in Israel. Water Policy 16, 79–101.

Gilmont, M., 2015. Water resource decoupling in the MENA through food trade as a mechanism for circumventing national water scarcity. Food Secur. 7, 1–19.

Gilron, J., Gara, N., Kedem, O., 2001. Experimental analysis of negative salt rejection in nanofiltration membranes. J. Membr. Sci. 185, 223–236.

Gonzales, E.J., Reinhardt, R.M., 1986. Foam finishing treatment of cotton fabrics with formaldehyde. Part I. Textile physical properties. Text. Res. J. 56, 497–502.

Gonzales, E.J., Reinhardt, R.M., 1987. Foam finishing treatment of cotton fabrics with formaldehyde. Part II. Formaldehyde properties. Text. Res. J. 57, 113–117.

Gregorian, R.S., Namboodri, C.G., Young, R.E., Baitinger, W.F., 1983. Foam application of phosphonium salt flame retardants. Text. Res. J. 53, 148–152.

Gu, Y.F., Dong, Y.N., Wang, H.T., Keller, A., Xu, J., Chiramba, T., Li, F.T., 2016. Quantification of the water, energy and carbon footprints of wastewater treatment plants in China considering a water-energy nexus perspective. Ecol. Indic. 60, 402–409.

Guo, H.C., Chen, B., Yu, X.L., Huang, G.H., Liu, L., Nie, X.H., 2006. Assessment of cleaner production options for alcohol industry of China: a study in the Shouguang Alcohol Factory. J. Clean. Prod. 14, 94–103.

Hasanbeigi, A., Price, L., 2015. A technical review of emerging technologies for energy and water efficiency and pollution reduction in the textile industry. J. Clean. Prod. 95, 30–44.

Hazrat, A., Mehtab, K., Muhammad, I., Sohail, A.J., 2013. Biological decolorization of crystal violet by *Alternaria solani*. Int. J. Green Herbal Chem. 2, 31–38.

Joa, B., Hottenroth, H., Jungmichel, N., Schmidt, M., 2014. Introduction of a feasible performance indicator for corporate water accounting—a case study on the cotton textile chain. J. Clean. Prod. 82, 143–153.

Juanjuan, L., Yongzhu, C., Lihua, L., Xiao, W., 2015. Water/oil repellent and foam finishing of polyester non-woven filter cloth processed by plasma. J. Dalian Polytech. Univ. 3.

Kiran-Ciliz, N., 2003. Reduction in resource consumption by process modifications in cotton wet processes. J. Clean. Prod. 11, 481–486.

Klose, G., 1993. Textile wastewater treatment without chemicals by the Klose-Clearox system. Dtsch. Faerber Kal. 97, 92–102.

Kocabas, A.M., Yukseler, H., Dilek, F.B., Yetis, U., 2009. Adoption of European Union's IPPC Directive to a textile mill: analysis of water and energy consumption. J. Environ. Manage. 91, 102–113.

Koyuncu, I., Topacik, D., 2002. Effect of organic ion on the separation of salts by nanofiltration membranes. J. Membr. Sci. 195, 247–263.

Koyuncu, I., Topacik, D., Yuksel, E., 2004. Reuse of reactive dyehouse wastewater by nanofiltration: process water quality and economical implications. Sep. Purif. Technol. 36, 77–85.

Laforest, V., 2014. Assessment of emerging and innovative techniques considering best available technique performances. Resour. Conserv. Recycl. 92, 11–24.

Lazarova, V., 2000. Wastewater disinfection: assessment of the available technologies for water reclamation. In: Goosen, M.F.A., Shayya, W.H. (Eds.), Water Conservation. Vol. 3. Water Management, Purification and Conservation in Arid Climates. Technomic.

Legislative Affairs Office of the State Council of the People's Republic of China, 2003. Cleaner Production Promotion Law of the People's Republic of China. Available from: http://www.chinalaw.gov.cn/article/fgkd/xfg/fl/201203/20120300360887.

Lei, Y.T., Su, L., 2015. Decoupling relationship between industrial water use and economic development in Guangdong Province. Water Conserv. Sci. Technol. Econ. 21, 1–4.

Levenstein, R., Hasson, D., Semiat, R., 1996. Utilization of the Donnan effect for improving electrolyte separation with nanofiltration membranes. J. Membr. Sci. 116, 77–92.

Li, K., Zhang, J., Gong, J., 2014. Wrinkle-resistant finish of foam technology for cotton fabric. J. Ind. Text. 43, 525–535.

Marinas, B.J., 1991. Reverse osmosis technology for wastewater reuse. Water Sci. Technol. 24, 215–227.

Marucci, M., Nosenzo, G., Capanelli, G., Ciabatti, I., Corrieri, D., Ciardelli, G., 2001. Treatment and reuse of textile effluents based on new ultrafiltration and other membrane technologies. Desalination 138, 75–82.

Mirjalili, M., Nazarpoor, K., Karimi, L., 2011. Eco-friendly dyeing of wool using natural dye from weld as co-partner with synthetic dye. J. Clean. Prod. 19 (9), 1045–1051.

Muntaha, S.T., Khan, M.N., 2015. Natural surfactant extracted from *Sapindus mukurossi* as an eco-friendly alternate to synthetic surfactant—a dye surfactant interaction study. J. Clean. Prod. 93, 145–150.

Nasr, L., Ayda, B., Abdessalem, S.B., 2016. Impact of modified DMDHEU and copolymer acrylic resin using spraying treatment before and after an enzymatic washing on the mechanical properties of denim cotton fabric. J. Polym. Text. Eng. 3, 24–34.

National Bureau of Statistics of the People's Republic of China, 2015. China's Environmental Statistics Report from 2000–2014. Available from: http://www.stats.gov.cn/ztjc/ztsj/hjtjzl.

Organization for Economic Co-operation and Development, 2001. Decoupling: A Conceptual Overview. OECD Publishing, Paris, France.

Organization for Economic Co-operation and Development (OECD), 2002. Indicators to Measure Decoupling of Environmental Pressure from Economic Growth. OECD Publishing, Paris, France.

Ozturk, E., Yetis, U., Dilek, F.B., Demirer, G.N., 2009. A chemical substitution study for a wet processing textile mill in Turkey. J. Clean. Prod. 17, 239–247.

Pan, A., Chen, L., 2014. Decoupling and water footprint analysis of the coordinate development between water utilization and the economy in Hubei. Resour. Sci. 36, 328–333.

Parvinzadeh, M., 2009. Ultrasonic assisted finishing of cotton with nonionic softener. Tenside Surfact. Deterg. 46 (6), 335–339.

Perry, M., Linder, C., 1989. Intermediate reverse osmosis ultrafiltration (RO/UF) membranes for concentration and desalting of low molecular weight organic solutes. Desalination 71, 233–245.

Piatti, R., Amoroso, G., Frangi, P., Fini, A., 2011. Evaluation of zero-run off irrigation systems for containerized nursery production. Int. Symp. High Technol. Greenhouse Syst. (GreenSys) 893, 1173–1177.

Ramachandran, T., Karthik, T., Saravanan, D., 2008. Novel trends in textile wet processing. J. Instit. Eng. 89, 3–10.

Rowland, S.P., Bertoniere, N.R., King, W.D., 1983. Durable press performance and reagent distribution from foam application of DMDHEU. Text. Res. J. 53, 197–204.

Shaikh, M.A., 2009. Water conservation in textile industry. Pak. Text. J. 48–51.

Spiegler, K.S., Kedem, O., 1966. Thermodynamics of hyperfiltration (RO): criteria for efficient membranes. Desalination 1, 311.

Takahashi, N., Kumagai, T., 2006. Removal of dissolved organic carbon and color from dyeing wastewater by pre-ozonation and subsequent biological treatment. Ozone Sci. Eng. 28, 199–205.

Vajnhandl, S., Valh, J.V., 2014. The status of water reuse in European textile sector. J. Environ. Manag. 141, 29–35.

Van der Bruggen, B., Daems, B., Wilms, D., Vandecasteele, C., 2001. Mechanisms of retention and flux decline for the nanofiltration of dye baths from textile industry. Sep. Purif. Technol. 22–23, 519–528.

Wadsworth, L.C., Wey, P., 1988. Effects of differential foam application of durable press and fluorochemical finishes to cotton fabric. J. Ind. Text. 17, 152–166.

Wang, L.L., Ding, X.M., Wu, X.Y., 2013b. Blue and grey water footprint of textile industry in China. Water Sci. Technol. 11, 2485–2490.

Wang, L.L., Ding, X.M., Wu, X.Y., Yu, J.M., 2013a. Textiles industrial water footprint: methodology and study. J. Sci. Ind. Res. 72, 710–715.

World Trade Organization, 2016. World Trade Organization (WTO) International Trade Statistics form 2000–2014. Available from: https://www.wto.org/english/res_e/statis_e/its_e.htm.

Wu, D., Wang, Y.H., 2013. Evaluation and prospect on the decoupling trend of economic development and water environment pressure in China. Resour. Environ. Yangtze Basin 22, 1103–1109.

Xu, X., Spencer, H.G., 1997. Dye-salt separation by nanofiltration using weak acid polyelectrolyte membranes. Desalination 114, 129–137.

Yan, Y., Jia, J., Wang, L., Du, C., Liu, X., Fu, X., Liu, X., Wu, G., 2014. Study of the industrial water footprint of several typical cotton textiles in China. Acta Ecol. Sin.. 23.

Zhang, B., Wang, L., Luo, L.F., King, M.W., 2014. Natural dye extracted from Chinese gall—the application of color and antibacterial activity to wool fabric. J. Clean. Prod. 80, 204–210.

Zhang, Y., Yang, Q.S., 2014. Decoupling agricultural water consumption and environmental impact from crop production based on the water footprint method: a case study of the Heilongjiang land reclamation area, China. Ecol. Indic. 43, 29–35.

Zhang, Y.Q., Wei, X.C., Long, J.J., 2016. Ecofriendly synthesis and application of special disperse reactive dyes in waterless coloration of wool with supercritical carbondioxide. J. Clean. Prod. 133, 746–756.

Zhu, B., Song, Q., Liu, J., Liu, J., Gao, W., Li, L., 2015. Effects of foaming parameters on sized-foam properties. Text. Res. J. https://doi.org/10.1177/0040517515621128.

Zhu, H., Li, W., Yu, J., Sun, W., Yao, X., 2013. An analysis of decoupling relationships of water uses and economic development in the two provinces of Yunnan and Guizhou during the first ten years of implementing the great western development strategy. Procedia Environ. Sci. 18, 864–870.

Further reading

Bergenthal, J.F. Wastewater Recycle and Reuse Potential for Indirect Discharge Textile Finishing Mills: Volume 1-Technical Report. U.S. Environmental Protection Agency.

Emrah, A., Goksel, N.D., 2015. Reducing water and energy consumption in chemical industry by sustainable production approach: a pilot study for polyethylene terephthalate production. J. Clean. Prod. 99, 119–128.

Environmental Pollution Control, 1978. Textile Processing Industry. U.S. Environmental Protection Agency, Cincinnati, OH.

Hoekstra, A.Y., Chapagain, A.K., Aldaya, M.M., Mekonnen, M.M., 2011. The Water Footprint Assessment Manual: Setting the Global Standard. Earthscan, London.

Ozturk, E., Koseoglu, H., Karaboyacı, M., Yigit, N.O., Yetis, U., Kitis, M., 2016. Minimization of water and chemical use in a cotton/polyester fabric dyeing textile mill. J. Clean. Prod. 130, 92–102.

Index

Note: Page numbers followed by *f* indicate figures and *t* indicate tables.

A

Advanced oxidation processes, 182–184
African fashion and luxury industry, 110–111
Air flow dyeing machines, 142–143
Alternative fibers, 107–108
Atmospheric pressure cold plasma, 48

B

Biochemical oxygen demand (BOD), 12–13
Biomagnification, 13
Biopolishing, 147

C

Chlorine, 14
Chromium-tanned leather, 108
Coating, 50–51
Cold pad batch (CPB) method
   for pretreatment process, 142
   reactive dyes, 142
Continuous ink jet printing, 49–50
Cotton, 98
Counter current washing techniques, 143, 143*f*

D

Decortication, 109
Degassing technique, 52–53
Denim, 97–98
Digital printing, 49–50, 170
Direct electrochemical dyeing process, 47–48
DReAM machine, 49–50
Drop-on-demand technology, 49–50
Dyeing machines
  air flow, 142–143
  and process developments
    airflow machine, 47
    CDR and CBR, 47
    electrochemical dyeing process, 47–48
    enzyme-based processing, 49
    high-speed drying machines, 47
    irradiation techniques, 48
    jigger machine, 47
    microwave machine, 47
    peracetic acid bleaching, 49
  soft flow, 142
Dyeing process, 49–50
  cone, 122, 124*f*, 126*t*, 126*f*, 127, 131*f*
  electrochemical, 47–48
  HT rope dyeing, 122–125, 123*f*, 125*f*, 125*t*
  plasma, 64–65, 66*f*, 168–169
  supercritical fluid-based, 51
  ultrasonic technique, 52–53
  using supercritical CO_2, 66–68, 67*f*
    (*see also* Supercritical carbon dioxide (SC CO_2) dyeing method)
  zero discharge, 148–149

E

Eco-friendly fashion, 105–106
Electrochemical dyeing process, 47–48
Enzyme-based processing, 49
Eutrophication, 12

F

Fabrics, 22–23
  in dying industry, 22–23
  production phase, 33–35
Fallacy of clean luxury, 100
Fiber, 22
  classification, 23*f*, 33
  definition, 22
  natural fibers, 33
  spinning technologies, 33
  synthetic fibers, 33
Flax fabric, 108
Foam finishing
  reduce water consumption, 50
  textile wet processing, 140–141, 141*f*

G

Global warming, 156
Green advertising campaign, 105–106

I

Indirect electrochemical dyeing process, 47–48
Integrated water resource management (IWRM) model, 117–119

K

Knitting process, 33

L

Leather, 108
Low wet pickup techniques
  expression technique
    air jet ejectors, 138, 138f
    dehydration systems, 137, 137f
    fiber-filled rollers, 136, 136f
    vacuum slot extractors, 137, 138f
  squeezing operation, 135–136
  topical techniques
    engraved roller system, 139, 139f
    kiss roller coating, 139, 140f
    loop transfer coating, 139, 140f
    spraying system, 140, 141f
  water imbibition value, 136
Luxury and fashion industry
  in Africa, 110–111
  alternative fibers, 107–108
  educate consumers, 101–103
  fast-fashion, 103–105
  green communication campaign, 105–106
  leather, 108
  low carbon economy, 110
  new fibers, 109
  organic cotton, 107
  sustainability, 99–101
Luxury consumers, 100

M

Millitron digital printing machine, 49–50

O

Organic cotton, 107
Osmosis, reverse, 150
Ozone layer depletion, 156

P

Pareto analysis technique, 121
  drop-fill process
    cone dyeing process, 127, 131f
    HT reactive dyeing process, 127, 128–129f
  overflow rinsing process
    cone dyeing process, 122, 124f, 126t, 126f
    HT rope dyeing, 122–125, 123f, 125f, 125t
Peracetic acid bleaching, 49
Pinatex, 109
Plasma dyeing, 64–65, 66f, 168–169
Plasma irradiation, 48, 53
Printing process, 49–50

R

Rapid Enzymatic Single-bath Treatment (REST), 118
Reduce water consumption
  coloration by printing process, 49–50
  dyeing machinery and process developments
    airflow machine, 47
    CDR and CBR, 47
    electrochemical dyeing process, 47–48
    enzyme-based processing, 49
    high-speed drying machines, 47
    irradiation techniques, 48
    jigger machine, 47
    microwave machine, 47
    peracetic acid bleaching, 49
  foam finishing, 50
  low-water/solvent-based coating/nanocoating, 50–51
  powder coating, 50–51
  spray technique, 50
  supercritical fluid-based dyeing process, 51
  ultrasonic technique, 52–53
Reverse osmosis, 150

S

Sewage, 13–14
Soft flow dyeing machines, 142
Spray technique, 50
Supercritical carbon dioxide (SC CO_2) dyeing method, 66–68, 67f
  textile wet processing
    biopolishing, 147
    disperse dyes, 144–145, 145t, 146f
    vs. dyeing in water, 147, 147t
    flow properties and reactivity, 144, 145t

natural fibers and blends, 145–147
silicone crosslinking, 147
waterless chemical processing
  vs. conventional dyeing, 167–168, 168*t*
  disadvantages, 167–168
  pressure-temperature *(P-T)* phase diagram, 167, 168*f*
Supercritical fluid-based dyeing process, 51
Sustainability
  definition, 156
  global warming, 156
  luxury and fashion industry, 99–101
  needs, 156
  ozone layer depletion, 156
  for water, 156–157
  water conservation
    advanced oxidation processes, 182–184
    foam application technology, 181–182
    membrane separation, 185–186
    nanofiltration (NF) process, 186–187
    pollution control method, 184–185
    process-level water conservation, 184–185
    water footprint method, 180–181
  WF in fashion industry, 78

T

Textile industry
  dyes and chemicals, 21
  effluent composition and chemical usage, 28–29, 29–30*t*
  fabric production phase, 33–35
  fabrics, 22–23
  finishing processes
    drying, 25
    dyeing, 24–25
    finishing, 25
    softening, 25
    washing, 25
  impacts and environmental effects, 35, 36*f*
  in India, 21–22
  pollution prevention techniques
    challenges, 35
    manufacturing level, 35
  pretreatment process, 24
  schematic representation, 22, 23*f*
  untreated effluents, 21
  water conservation, 37
  water consumption
    high-quality water, 26
    humankind and environment, 25–26
    impact of water resources, 27
    low-quality water, 26
    moderate-quality water, 26
    pollution, 25–26
    quality characterization, 25–26
    standards, 69–70
    wastewater characteristics, 28–29
  water footprint management
    challenges and research opportunities, 86–87
    in China, 82
    on freshwater resources, 83, 83*f*
    metal zipper, 83
    in Spain, 82
    in Uzbekistan, 82–83
  water quality characteristics and standards, 27, 28*t*
  water reuse and recycling
    end-of-pipe approach, 38
    start-of-pipe approach, 37
  yarn production phase
    knitting process, 33
    wet treatment, 34–35
Textile product manufacturing process, 69–70, 70*f*
Textile wet processing, water conservation, 63–68
  air flow dyeing machines, 142–143
  cold pad batch method
    for pretreatment process, 142
    reactive dyes, 142
  counter current washing techniques, 143, 143*f*
  decolorizing methods for dyes, 149, 149*t*
  foam finishing, 140–141, 141*f*
  low wet pickup techniques
    expression technique, 136–138
    topical techniques, 138–140
  soft flow dyeing machines, 142
  supercritical carbon dioxide (SC-CO_2)
    biopolishing, 147
    disperse dyes, 144–145, 145*t*, 146*f*
    vs. dyeing in water, 147, 147*t*
    flow properties and reactivity, 144, 145*t*
    natural fibers and blends, 145–147
    silicone crosslinking, 147
True color, 49–50

U

Ultraviolet (UV) irradiation, 48, 53

V

Vegetable-tanned leather, 108
Virtual water, 17
Viscose, 107

W

Wastewater
　characteristics, 28–29
　issues, 61–63, 62f
　jet weaving wastewater, 178
　soaper wastewater, 178
　treatment, 13–14
　discharge reduction, in textile, 118
Water (H_2O)
　challenges, 17–18
　characteristics, 5
　in food and beverages, 1–2
　groundwater, 3
　in human body, 1–2
　hydropower generation, 1–2
　in industry, 1–2
　pollution
　　anthropogenic/artificial sources, 12
　　biochemical oxygen demand, 12–13
　　biomagnification, 13
　　bug sprays, 9–11
　　by businesses, 9–11
　　consequences, 11
　　contamination control, 13
　　eutrophication, 12
　　factors determining, 11
　　industrialization and overpopulation, 9–11
　　nonpoint sources, 12
　　pesticides and pharmaceuticals, 9–11
　　point sources, 12
　　primary treatment, 14
　　sanitation, 9–11
　　secondary treatment, 14
　　tertiary treatment, 14
　　urbanization, 9–11
　at room temperature, 1–2
　structure, 3–4, 4f
　surface water
　　benefits, 3
　　challenges, 3
　　provincial territories, 2–3
　　runoff, 2–3
　unique aspects, 3–4

universal solvent, 3–4
water consumption, 5–6
water shortages
　in Asia, 16
　causes, 15–16
　reasons behind, 15
　solutions, 16–17
water use
　data analysis and management, 8
　for fibers, 32t
　irrigators, 6
　offstream and instream water, 5–6
　open water providers, 6
　preservation programs, 6
　primary data collection, 6–7
　secondary data collection, 7–8
water withdrawals, 9
Water availability in India
　classification, 157–158
　process-wise water consumption, 160–161, 161t
　segments, 157
　textile sector contribution, 157
　useable quantity, 157
　water consumption, 158, 158f
　water requirement
　　cotton textile processing industry, 159–160, 160t
　　for Indian cotton sector, 161, 162t
　　for textile processing industry, 158–159, 159–160t
Water conservation, 63–68
　challenges and opportunities, 188–189
　effective dyeing machines, 164–165
　finishing processes, 176
　good housekeeping measures, 165–167
　measures, 177–178
　payback period, 177–178
　process and chemical management
　　alkaline hydrogen peroxide, 163
　　bleaching, 162–163
　　desizing, 162
　　dyeing, 163
　　material-to-liquor ratio, 163, 164t
　　percentages of unfixed dyes, 163, 163t
　　scouring, 162
　reuse programs
　　bleach bath cotton, 178
　　cooling water, 179
　　dye liquor, 179

final rinse water from dyeing, 178
jet weaving wastewater, 178
soaper wastewater, 178
wash water, 179–180
sustainable strategies
advanced oxidation processes, 182–184
foam application technology, 181–182
membrane separation, 185–186
nanofiltration (NF) process, 186–187
pollution control method, 184–185
process-level water conservation, 184–185
water footprint method, 180–181
textile industry, 37
in textile wet processing
air flow dyeing machines, 142–143
cold pad batch method, 142
counter current washing techniques, 143, 143*f*
decolorizing methods for dyes, 149, 149*t*
foam finishing, 140–141, 141*f*
low wet pickup techniques, 135–140
soft flow dyeing machines, 142
supercritical carbon dioxide (SC-CO$_2$), 144–147
Water consumption, 5–6
in Bangladesh, 120
effluent generation
air and water pollution, 54
optimizing water consumption, 148
reverse osmosis, 150
and treatment, 55
use of reactive dyes, 55
waste minimization, 55
zero discharge dyeing, 148–149
in India, 158, 158*f*
in mechanical processing, 43
in processing natural fibers, 44–45, 44–45*t*
in processing synthetic fibers, 45, 46*t*
reduction strategies
coloration by printing process, 49–50
dyeing machinery and process developments, 47–49
foam finishing, 50
low-water/solvent-based coating/nanocoating, 50–51
powder coating, 50–51
spray technique, 50
supercritical fluid-based dyeing process, 51

ultrasonic technique, 52–53
textile industry
high-quality water, 26
humankind and environment, 25–26
impact of water resources, 27
low-quality water, 26
moderate-quality water, 26
pollution, 25–26
quality characterization, 25–26
standards, 69–70
wastewater characteristics, 28–29
in Turkey
environmental impacts, 119, 119*t*
Environmental Performance Evaluation study, 119, 119*f*
in HT reactive dyeing, 122–125, 125*t*
industrial fresh water costs, 116–117, 117*f*
Pareto analysis technique, 121
ratio for cotton wet operations, 116, 116*f*
reduced water consumption, 127–132
reduction of salt consumption, 127–129
seven-stage approach, 120
textile dye house processes, 116
textile wastewater discharge reduction, 118
water saving and pollution prevention models, 117
ultraviolet and plasma irradiation, 53
in wet processing, 43, 43*t*
Water demand, 61–64
Water footprint (WF)
clothing industry, 78
in fashion industry
in China, 97
cotton, 98
denim, 97–98
emerging macro-level trends, 84–86
environmental impact, 81
sustainability, 78, 95
vs. textile industry, 99
fashion supply chains
disclosure framework, 79–80
framework, 81, 81*f*
in Hong Kong, 80–81
Indian textile industry, 79–80
Italian fashion companies, 80–81
in Malaysia, 80–81
progressive transition drivers, 79–80
societal and environmental issues, 79–80

Water footprint (WF) *(Continued)*
  in South Africa, 80–81
  in Sri Lanka, 80–81
  taxation schemes, 81
  green, blue, and gray WFs, 77–78, 95–96
  life cycle analysis method, 77–78
  luxury and fashion market
    in Africa, 110–111
    alternative fibers, 107–108
    educate consumers, 101–103
    fast-fashion, 103–105
    green communication campaign, 105–106
    leather, 108
    low carbon economy, 110
    new fibers, 109
    organic cotton, 107
    sustainability, 99–101
  sanitation, 96
  in textile industry
    challenges and research opportunities, 86–87
    in China, 82
    on freshwater resources, 83, 83f
    metal zipper, 83
    in Spain, 82
    in Uzbekistan, 82–83
Waterless chemical processing
  digital textile printing, 170
  plasma processing, 168–169
  supercritical carbon dioxide-based processing
    vs. conventional dyeing, 167–168, 168t
    disadvantages, 167–168
    pressure-temperature *(P-T)* phase diagram, 167, 168f
  zero liquid discharge, 169–170
Water management model, 117
Water pollution
  anthropogenic/artificial sources, 12
  biochemical oxygen demand, 12–13
  biomagnification, 13
  bug sprays, 9–11
  by businesses, 9–11
  consequences, 11
  contamination control, 13
  eutrophication, 12
  factors determining, 11
  industrialization and overpopulation, 9–11
  nonpoint sources, 12
  pesticides and pharmaceuticals, 9–11
  point sources, 12
  primary treatment, 14
  sanitation, 9–11
  secondary treatment, 14
  tertiary treatment, 14
  textile industry, 25–26
    challenges, 35
    manufacturing level, 35
  urbanization, 9–11
Water requirement
  cotton textile processing industry, 159–160, 160t
  for Indian cotton sector, 161, 162t
  for textile processing industry, 158–159, 159–160t
Water reuse and recycling, 188–189
  bleach bath cotton, 178
  cooling water, 179
  dye liquor, 179
  final rinse water from dyeing, 178
  jet weaving wastewater, 178
  soaper wastewater, 178
  textile industry
    end-of-pipe approach, 38
    start-of-pipe approach, 37
  wash water, 179–180
Water withdrawals, 9
  in European Union, 63–64
  in India and China, 63–64
  laws and regulations, 68–69
  opportunities and limitations
    washing and flushing, 71
    water reduction steps, 71
  plasma dyeing, 64–65, 66f
  standards, 69–70
  supercritical carbon dioxide dyeing method, 66–68, 67f
  in Turkey, 63–64
  wet and dry processes, 61–63, 63t

Z

Zero liquid discharge, 169–170